APPARENT PLACES OF FUNDAMENTAL STARS 2025

for 64 stars selected from the
Sixth Catalogue of Fundamental Stars

ASTRONOMISCHES RECHEN-INSTITUT
Zentrum für Astronomie der Universität Heidelberg

Apparent Places of Fundamental Stars 2025

URL of the journal: https://journals.ub.uni-heidelberg.de/index.php/apfs

Editor: Astronomisches Rechen-Institut, Zentrum für Astronomie der Universität Heidelberg, Mönchhofstraße 12–14, 69120 Heidelberg, Telefon: (06221) 54-1850

Responsible for the contents: Dr. R. W. Schmidt
Prof. Dr. J. Wambsganß

UNIVERSITÄTS-
BIBLIOTHEK
HEIDELBERG

Published by heiJOURNALS, 2025

Universität Heidelberg / Universitätsbibliothek
heiJOURNALS
Grabengasse 1, 69117 Heidelberg
https://journals.ub.uni-heidelberg.de
E-Mail: ub@ub.uni-heidelberg.de

The online version of this publication is hosted permanently and freely available (open access) by heiJOURNALS, the e-Journal platform of Heidelberg University Library, https://journals.ub.uni-heidelberg.de.
doi: https://doi.org/10.60653/apfs.2025

ISSN 0174-254X
eISSN 2943-8004

ISBN 978-3-947733-15-6 (Print)
ISBN 978-3-947733-14-9 (PDF)

PREFACE

This booklet, published annually, is the continuation of the series of volumes "Apparent Places of Fundamental Stars", (APFS) which was first published in 1941. The compilation and publication of the first nineteen volumes was undertaken by H. M. Nautical Office, Royal Greenwich Observatory. In accordance with a recommendation of the I.A.U. (Dublin meeting 1955) this task was taken over by the Astronomisches Rechen-Institut, Heidelberg, from the twentieth volume onwards. Starting with the edition of the year 2000, the extensively large books were replaced by the present small booklet. Starting in 2006 only the introductory remarks concerning the CIO-based (celestial intermediate origin) and the equinox-based method are provided. The apparent places for 64 stars reduced with the equinox-based method are given in the printed version; the data for 878 fundamental stars (FK6) and for Polaris are provided via the Internet in consideration of both methods (http://www.ari.uni-heidelberg.de/ariapfs). IAU 2000/2006 precession-nutation is used for intermediate and apparent positions. The underlying precession-nutation model is recommended by IAU 2006 Resolution B1.

Since the year 2000 the apparent and mean places have been based on the so-called single-star solution of the Sixth Catalogue of Fundamental Stars (FK6, Wielen et al., Veröff. Astron. Rechen-Institut, Heidelberg, No. 35, 1999). This catalogue is on the astrometric system defined by the HIPPARCOS catalogue (ESA SP-1200, 1997), which has been adopted as the primary celestial reference frame at optical wavelengths since 1998 (*Trans. IAU,* **23B**, 39, 1999). In FK6 we have improved the HIPPARCOS proper motions by combining the HIPPARCOS observations with the data given in FK5.

In the online version we provide the intermediate and apparent places reduced by the *CIO-based* and by the *equinox-based* method, which differ only in right ascension. The entries corresponding to sidereal days divisible by ten (corresponding in the printed volumes) are marked by an asterisk.

Intermediate and apparent places can also be obtained conveniently within the framework of the "German Astrophysical Virtual Observatory" (GAVO). The corresponding web page is accessible via the ARI APFS web pages or directly by http://vo.uni-hd.de/apfs. Apparent and intermediate places for 878 stars of the FK6 Part I and 3272 stars of the FK6 Part III can be retrieved conveniently. This service is completed by Hipparcos-based apparent and intermediate places of additional stars, and by using data from Gaia Data Release 3 (https://www.cosmos.esa.int/web/gaia/dr3; Gaia Data Release 3: Summary of the contents and survey properties, Gaia Collaboration (Vallenari, A., Brown, A. G. A., et al. 2023, A&A 674, A1). A facility to compute the Earth Rotation Angle (ERA), Greenwich Apparent Sidereal Time (GAST) and Greenwich Mean Sidereal Time (GMST) is available on a related web page. Some introductory remarks and references can be found in the corresponding service info. Any comments would be appreciated.

R. W. SCHMIDT E. K. GREBEL J. WAMBSGANSS

Heidelberg, October 2024
Astronomisches Rechen-Institut am Zentrum für Astronomie der Universität Heidelberg

---

Preface
Published under CC BY 4.0, doi: https://doi.org/10.60653/apfs.2025

CONTENTS

INTRODUCTION

This booklet provides examples for the reduction of 64 mean star places to apparent places for the year 2025. In view of the high accuracy of the mean positions and proper motions in the FK6 and in the HIPPARCOS catalogue we have increased the number of significant digits in the tabulated apparent places by one compared with the volumes before 2000. Since no "short period terms" are included in the apparent places following 2005 there exists no interpolation to intermediate dates based on the printed version. The internet version listing "daily" apparent places should be used for an approximate interpolation.

In this booklet we present apparent places for only a few fundamental stars as examples. The intermediate and apparent places for a large number of fundamental stars are provided by the Astronomisches Rechen-Institut at the URL:

http://www.ari.uni-heidelberg.de/ariapfs

We tabulate the intermediate and apparent places for *daily* upper culminations at Greenwich. Those entries that would have been published according to the conventions of the printed volumes are marked by an asterisk in the online version. It may be noted here that we can easily use the program for transits over other meridians just by subtracting the longitude of an observer from the Greenwich sidereal date of the transit.

At its 23rd General Assembly in 1997, the International Astronomical Union decided (*Trans. IAU* **23B** 39, 1999) to adopt an International Celestial Reference System (ICRS). The ICRS is realized at optical wavelengths by stars in the Hipparcos Catalogue, in particular by those having reliable proper motions. This subset - comprising more than 80 per cent of the stars of the Hipparcos Catalogue - constitutes the Hipparcos Celestial Reference Frame (HCRF).

The new highly accurate IERS-observing technique (VLBI) has recommended to adopt a new zero point for the equatorial system. Guinot's non-rotating origin (Guinot, B., 1979, In: McCarthy, D.D., Pilkington, J.D. (eds.), Time and the Earth's Rotation. D. Reidel Publ. Co., p. 7) was adopted for substituting the classical equinox. This origin is stable in such a way that there are merely motions of the new zero point in right angles to the instantaneous equator. With this new definition the rotation of the Earth is given directly as the difference between the non-rotating origin and the terrestrial origin. This difference is directly proportional to UT1 and no precession-nutation terms are included.

In addition to the CIO-based procedure we also give the apparent places using the classical equinox as the origin in right ascension; this older method may still be used in many applications. No differences in declination occur since the equator remains unchanged.

Precession and nutation reductions agree with IAU2000/2006 precession-nutation in accordance with IAU 2006 Resolution B1 (see e.g. Capitaine & Wallace, A&A 450, 855 (2006)).

Software Routines from the IAU SOFA Collection were used. Copyright © International Astronomical Union Standards of Fundamental Astronomy (http://www.iausofa.org). DE430/LE430 ephemerides (ftp://ssd.jpl.nasa.gov/pub/eph/planets/README.txt) are used for GAVO.

Introduction

From VLBI observations it has been found that there exists a "celestial pole offset" between a CIO-based and a J2000 right ascension that has to be applied before making use of the precession-nutation terms. This bias is described more explicitly in Feissel, M., Mignard, F. (A&A 331, L33 (1998)) and in Hilton, J.L., Hohenkerk, C.Y. (A&A 413, 765 (2004)). It is, however, included already in the SOFA subroutines, and no changes to the input files were applied.

The input data are the HIPPARCOS and the FK6 catalogue. The parallaxes used are those from the HIPPARCOS catalogues, and the radial velocities are taken from the machine readable version of the FK6.

In our work on the FK6 we have shown that the HIPPARCOS proper motions are "instantaneous motions" that may differ in many cases significantly from the mean (centre of mass) motion. With the combination of the HIPPARCOS data and the FK5 catalogue (reduced to the HIPPARCOS system) we have derived proper motions in the FK6 which describe much better the mean motion. As part of our reduction process we have also obtained additional information on possible double stars as well as on stars that can be regarded with high probability as single star candidates. Single star candidates are best suited to maintain the International Reference System, and from the year 2000 onwards we only give apparent places for the single star candidates (except Polaris, see p. 7-8). For comparison we provide the differences between the given SI-solution with respect to the HIPPARCOS data and to the long-term prediction (LTP) in the FK6. This table is given on pages 34-35.

Apparent Places of 10-Day Stars (Pages 11-25)

Examples for the apparent places of the stars with declinations between ±81° are given for every tenth upper transit at Greenwich on pages 11-25. The choice of the data is fixed by the moment for which the integral part of the Greenwich sidereal date is divisible by 10. In this booklet we give the equinox-based right ascension, which has a much larger difference from date to date than those determined with the CIO-based method. The CIO-based method is given additionally in the internet version.

The column U.T. gives the approximate time of transit for the first star on the page; it is rounded to the nearest tenth of a day. For transits over other meridians the column U.T. can be regarded as the "local" mean solar date for that transit. For transits of other stars on that page the right ascension difference of the star from the first star should be taken into account.

The right ascension and declination are referred to the true equator and equinox, or non-rotating origin in the CIO-based method. Since the equator does not change merely the right ascensions are affected and the declinations remain unchanged.

From 1984 onwards (see Preface to APFS 1984) the mean positions of the stars in the FK6 or the HIPPARCOS catalogue are freed from the term of elliptic aberration (E-terms), which depends on the eccentricity of the Earth's orbit. This term is now included in the reduction to the apparent place.

The hours and minutes of right ascension and the degrees and minutes of declination given at the head of the columns are adjusted so that the seconds never change sign, though this may involve their exceeding 60.

Immediately below the tabulated right ascension and declination we provide:

(I) the mean place for the middle of the year.
(II) sec δ and tan δ corresponding to the mean place.
(III) the day upon which the star transits twice in upper culmination.

In the volumes prior to 2006, examples were given for interpolating the apparent place for other meridians, including in particular the short period terms of nutation. In the present booklet we do not to give any interpolation and the user should enter his or her subroutine using the appropriate time. An approximate transit could also be obtained from the daily positions given in the internet version.

Apparent Places of Circumpolar Stars (Pages 26-33)

Examples for the apparent places of circumpolar stars with declinations exceeding $\pm 81°$ are given for every upper transit at Greenwich on pages 26-33. Polaris (HIPPARCOS No. 11767, FK6 No. 907) is not included in the FK6 Part I, because it is a binary. The apparent places given on pages 26 and 27 are based on the retrograde orbit derived by Wielen et al. (A&A **360**, 399, (2000)). Each two facing pages are devoted to a star. In the left hand column only the day of the month is given without the fraction of the day. The right ascension and declination are referred to the true equator and equinox (and additionally the non-rotating origin in the Internet version), short period terms of nutation are included. Three decimals of a second are only given for the right ascensions. On the one day during the year when there are two upper transits at Greenwich both are shown.

The values of sec δ and tan δ are given for every month and refer to the apparent place on the 16th day of the month. The footnotes, repeated on each page, give the mean right ascension and declination in the middle of the year and the date of double lower transit.

Reduction to the HIPPARCOS catalogue
Reduction to the FK6 (Long-term prediction)
(Pages 34-35)

The FK6 is the result of combining the FK5 with the HIPPARCOS observations. Various solutions have been derived in the FK6, adopting different models for the star's kinematic behaviour, which leads to different weighting schemes in the least squares solutions. Two of these solutions are briefly mentioned here. For details please refer to the FK6.

The apparent places given in this booklet and in the internet version are based on the so-called single-star solution (abbreviated as SI) of the FK6 assuming that the star can be treated as a single star. Stars with no indication of a binary nature are best suited to maintain the International Celestial Reference System (ICRS). We have therefore restricted the stellar sample in this booklet mainly to such candidates, which are called "astrometrically excellent stars of the highest rank (***)" in the FK6. A few non-excellent stars are also included because of their brightness, their large foreshortening effects or their special importance (Polaris, a binary not included in the FK6, Part 1). Starting in 2006 we added ten other excellent stars, including two circumpolar stars.

In the FK6 long-term prediction (LTP) we admit a possible (but still undetected) binary nature of a star. In this case HIPPARCOS observed, more or less, an instantaneous proper motion, depending on the star's observational period. Combining the HIPPARCOS observations with highly-weighted FK5 data yields an FK6 proper motion that describes much better the star's position for epochs differing significantly from the HIPPARCOS epoch, about 1991.25. Examples can be found in the table on pages 34-35.

The table on pages 34-35 provides the differences of the FK6 single-star solution (SI) with the HIPPARCOS data on one hand, and with the long-term prediction in the FK6 (LTP) on the other hand. The data in the table hold for the middle of the year. Columns one and two list the FK6 and HIPPARCOS number, column three gives the difference in right ascension between the HIPPARCOS data and the single-star solution of the FK6 in units of 0.0001 seconds of time, column four is the proper motion difference in 0.0001 seconds of time per year, and columns five and six list the corresponding differences in declination in 0.001 arcsec and 0.001 arcsec/year, respectively. The differences between the long-term predicition and the single-star solutions in the FK6 are given similarly in the columns seven through ten. Polaris (FK6 No. 907) is a binary and needs a special treatment. The small table at the end of page 35 provides the data to reduce the apparent place of Polaris from the FK6 (p. 26-27) to the HIPPARCOS catalogue.

These data permit, for any date in the year, the computation of the corrections that have to be added to the tabulated apparent places in order to get the positions based on the HIPPARCOS catalogue or on the long-term prediction in the FK6 from the SI position.

Table UT–ST (Pages 36-39)
Sidereal Time at 0^{h} *U.T.*

On these pages are given in order of 0^{h} U.T. on each day of the year:

(I) the apparent (or true) sidereal time to $0\overset{\text{s}}{.}001$
(II) the mean (or uniform) sidereal time, given as seconds and decimals only, the hours and minutes being the same as in the first column
(III) difference Apparent – Mean (app–mean) in units of $0\overset{\text{s}}{.}001$

In the APFS volumes preceding 1960, the equation of equinoxes was designated as the nutation in right ascension.

Conversion of Mean Solar Time to Sidereal Time
Conversion of Sidereal Time to Mean Solar Time

The following relations derived from the expressions between mean solar time and mean sidereal time as given in *Trans. I.A.U.* **18B,** 72 (1983) are used. Both relations can be used in the following two examples.

$$1 \text{ mean solar day} = 24^{\text{h}}03^{\text{m}}56\overset{\text{s}}{.}55536\,79 \quad \text{in mean sidereal time} \qquad (1a)$$
$$1 \text{ mean sidereal day} = 23^{\text{h}}56^{\text{m}}04\overset{\text{s}}{.}09053\,08 \quad \text{in mean solar time} \qquad (1b)$$

The time dependence of these relations has been ignored, since it is of no practical importance in the present development.

In using the above relations for passing from mean solar time to apparent sidereal time and vice versa, we also must use the apparent sidereal time at 0^{h} taken from the last table on page 36-39. It must be remembered that a correction should be applied for the change of the equation of equinoxes between 0^h and the given U.T.

Thus the local apparent sidereal time at Heidelberg (Longitude = $0^h34^m53\overset{s}{.}190$) at U.T. $7^h\,21^m\,36\overset{s}{.}572$ on 2025 January 26 is obtained as:

Mean solar interval at 0^h		7^h	21^m	$36\overset{s}{.}572$
Correction to mean solar time to given sidereal time } (relation (1a))	+		1	12.445
	+			0.100
Apparent sidereal time at 0^h (Table p. 36)		8	22	09.848
Change in the equation of equinoxes from 0^h to 7^h (Table p. 36)	+			0.003
Sum = Greenwich apparent sidereal time		15	44	58.968
Longitude Heidelberg	+	0	34	53.190
Sum = Heidelberg apparent sidereal time		16	19	52.158

Similarly the U.T. on 2025, January 26 corresponding to an apparent sidereal time at Heidelberg of $16^h19^m52\overset{s}{.}158$ is obtained as:

Heidelberg apparent sidereal time		16^h	19^m	$52\overset{s}{.}158$
Longitude Heidelberg	−	0	34	53.190
Difference = Greenwich apparent sidereal time		15	44	58.968
Apparent sidereal time at 0^h (Table p. 36)		8	22	09.848
Sidereal interval		7	22	49.120
Correction to sidereal time to given mean solar time } (relation (1b))	−		1	12.445
	−			0.100
Change in the equation of equinoxes from 7^h to 0^h (Table p. 36)	−			0.003
Sum = required U.T.		7	21	36.572

Apparent places for different longitudes

Suppose the apparent position needs to be obtained for a star at upper transit in Heidelberg ($\lambda = +0^h\ 34^m53\overset{s}{.}190$, i.e. the star culminates 0.581 hours or 0.0242 days earlier than in Greenwich). We would use the apparent place subroutine for this, using as input date the Heidelberg transit which is 0.581 hours (or 0.0242 days) earlier than the corresponding transit in Greenwich. This information can be directly included in the subroutine.

For example the apparent position of 94 Piscium (FK6 No. 1039) may be required at upper transit in Heidelberg for 13th May 2025. We include the longitude of Heidelberg into the subroutine and we find an apparent position of

$\alpha = 1^h\,28^m\,02\overset{s}{.}8313$ (EQUINOX-based)

$\alpha = 1^h\,26^m\,44\overset{s}{.}8231$ (CIO-based)

$\delta = 19^\circ\,22'\,11\overset{''}{.}054$ (CIO-based = EQUINOX-based)

Diurnal Aberration

The diurnal aberration must be added to the right ascension for upper transits. Alternatively, it can be subtracted from the time of transit. In the case of lower transits the sign of the correction has to be reversed. With declination δ and latitude ϕ the given correction is

$$\text{Diurnal Aberration} = 0\overset{s}{.}0213 \cos\phi \sec\delta$$

A remark concerning the history of the APFS at ARI

Many staff members have contributed to the APFS since the first volume for 1960 was published. We would like to mention in particular F. Gondolatsch, T. Lederle, H. Schwan and H. Lenhardt in addition to the former directors W. Fricke and R. Wielen.

APPARENT PLACES OF STARS, 2025

TEN-DAY STARS AT UPPER TRANSIT AT GREENWICH

EQUINOX BASED RIGHT ASCENSION – WHOLE NUTATION

FK5-No.	10		22		44		1039	
HIP-No.	1599		3419		5661		6732	
Name	ζ Tucanae		β Ceti		102 G. Sculptoris		94 Piscium	
U.T.	R.A.	Dec.	R.A.	Dec.	R.A.	Dec.	R.A.	Dec.
	0^{h}21^{m}	**– 64°43′**	**0^{h}44^{m}**	**– 17°50′**	**1^{h}13^{m}**	**– 37°42′**	**1^{h}28^{m}**	**+ 19°22′**
12 27^{d}.7	22^{s}.4893	60$''$.029	50^{s}.8325	65$''$.279	55^{s}.0853	99$''$.418	03^{s}.1514	17$''$.632
1 6.7	22.1299	59.338	50.7286	65.730	54.9312	99.873	03.0479	17.339
1 16.7	21.7805	57.981	50.6385	65.856	54.7798	99.842	02.9581	16.934
1 26.7	21.4025	56.416	50.5040	66.057	54.5818	99.712	02.8149	16.073
2 5.6	21.1120	54.253	50.3921	65.879	54.4109	98.998	02.6737	15.249
2 15.6	20.9064	51.497	50.3223	65.255	54.2818	97.695	02.5673	14.543
2 25.6	20.7118	48.441	50.2604	64.462	54.1541	96.133	02.4763	13.670
3 7.6	20.5792	45.172	50.2091	63.491	54.0419	94.280	02.3880	12.733
3 17.5	20.5455	41.751	50.1841	62.291	53.9676	92.109	02.3136	11.906
3 27.5	20.6058	37.930	50.2296	60.612	53.9621	89.447	02.3194	11.397
4 6.5	20.7139	34.095	50.3050	58.764	53.9872	86.641	02.3656	10.958
4 16.4	20.8746	30.600	50.3774	57.008	54.0227	83.940	02.4005	10.492
4 26.4	21.1754	26.908	50.5328	54.793	54.1506	80.810	02.5174	10.519
5 6.4	21.5512	23.228	50.7533	52.298	54.3449	77.508	02.7177	10.939
5 16.4	21.9299	20.116	50.9636	50.049	54.5372	74.562	02.9158	11.318
5 26.3	22.4069	17.254	51.2145	47.688	54.7862	71.585	03.1482	12.076
6 5.3	22.9781	14.572	51.5271	45.125	55.1070	68.536	03.4457	13.340
6 15.3	23.5441	12.369	51.8513	42.711	55.4420	65.806	03.7713	14.733
6 25.3	24.1190	10.752	52.1704	40.536	55.7829	63.453	04.0948	16.223
7 5.2	24.7304	09.724	52.4900	38.571	56.1410	61.438	04.4111	17.927
7 15.2	25.3568	09.007	52.8382	36.631	56.5310	59.609	04.7657	19.949
7 25.2	25.9287	08.865	53.1702	35.009	56.9049	58.276	05.1198	21.966
8 4.1	26.4285	09.600	53.4378	34.008	57.2250	57.698	05.4072	23.761
8 14.1	26.9300	10.642	53.7105	33.124	57.5564	57.349	05.6972	25.781
8 24.1	27.3733	11.996	53.9801	32.423	57.8791	57.337	05.9996	27.884
9 3.1	27.6717	14.064	54.1720	32.374	58.1198	58.110	06.2366	29.580
9 13.0	27.9105	16.472	54.3260	32.644	58.3246	59.262	06.4311	31.190
9 23.0	28.1010	19.008	54.4670	33.058	58.5132	60.626	06.6135	32.844
10 3.0	28.1651	21.681	54.5709	33.753	58.6490	62.349	06.7760	34.273
10 13.0	28.1197	24.491	54.6226	34.764	58.7235	64.404	06.8902	35.405
10 22.9	27.9995	27.400	54.6213	36.062	58.7433	66.711	06.9409	36.290
11 1.9	27.8346	29.862	54.6265	37.212	58.7555	68.829	07.0008	37.214
11 11.9	27.5791	31.993	54.6101	38.374	58.7278	70.914	07.0486	37.919
11 21.8	27.2199	34.108	54.5184	39.865	58.6186	73.216	07.0117	38.117
12 1.8	26.8741	35.579	54.4324	41.087	58.5102	75.082	06.9631	38.340
12 11.8	26.5218	36.279	54.3697	41.938	58.4094	76.447	06.9388	38.588
12 21.8	26.0898	36.676	54.2556	42.898	58.2460	77.765	06.8598	38.359
12 31.7	25.6703	36.552	54.1220	43.709	58.0625	78.716	06.7402	37.920
Mean Place	22.9542	30.554	52.0933	49.367	55.7306	78.994	04.6300	18.395
sec δ, tan δ	+2.342	–2.118	+1.051	–0.322	+1.264	–0.773	+1.060	+0.352
Dble. Trans.	September 26		October 2		October 9		October 13	

Apparent places of ten-day stars
Published under CC BY 4.0, doi: https://doi.org/10.60653/apfs.2025

APPARENT PLACES OF STARS, 2025

TEN-DAY STARS AT UPPER TRANSIT AT GREENWICH

EQUINOX BASED RIGHT ASCENSION – WHOLE NUTATION

FK5-No.	1045		80		110		119	
HIP-No.	7513		10642		14240		15510	
Name	υ Andromedae		67 Ceti		μ Hor		92 G. Eridani	
U.T.	R.A.	Dec.	R.A.	Dec.	R.A.	Dec.	R.A.	Dec.
	$\mathbf{1^{h}38^{m}}$	$\mathbf{+\ 41^{\circ}31'}$	$\mathbf{2^{h}18^{m}}$	$\mathbf{-\ 6^{\circ}18'}$	$\mathbf{3^{h}04^{m}}$	$\mathbf{-\ 59^{\circ}37'}$	$\mathbf{3^{h}20^{m}}$	$\mathbf{-\ 42^{\circ}57'}$
12 $27.^{d}8$	$16.^{s}5759$	$61.''957$	$14.^{s}6232$	$27.''404$	$14.^{s}7720$	$97.''876$	$57.^{s}1365$	$96.''580$
1 6.8	16.4183	62.197	14.5405	28.126	14.4806	99.422	56.9975	98.248
1 16.7	16.2777	62.090	14.4574	28.713	14.1433	100.491	56.8309	99.545
1 26.7	16.0790	61.311	14.3145	29.462	13.7446	101.319	56.6003	100.676
2 5.7	15.8732	60.334	14.1711	29.836	13.3493	101.334	56.3631	101.066
2 15.7	15.7052	59.253	14.0529	29.838	12.9686	100.621	56.1374	100.793
2 25.6	15.5657	57.787	13.9344	29.827	12.5692	99.598	55.8959	100.274
3 7.6	15.4318	56.055	13.8124	29.652	12.1852	98.075	55.6522	99.287
3 17.6	15.3119	54.285	13.7034	29.171	11.8512	95.964	55.4311	97.715
3 27.6	15.2900	52.685	13.6609	28.310	11.5700	93.326	55.2612	95.640
4 6.5	15.3268	51.028	13.6473	27.355	11.3259	90.474	55.1174	93.352
4 16.5	15.3576	49.293	13.6268	26.373	11.1334	87.484	54.9903	90.873
4 26.5	15.4787	48.068	13.6856	24.847	11.0422	83.935	54.9459	87.805
5 6.4	15.7027	47.206	13.8170	23.055	11.0273	80.228	54.9701	84.541
5 16.4	15.9362	46.342	13.9477	21.428	11.0554	76.785	55.0136	81.468
5 26.4	16.2058	45.933	14.1215	19.503	11.1782	73.144	55.1220	78.117
6 5.4	16.5493	46.141	14.3649	17.245	11.4015	69.381	55.3121	74.563
6 15.3	16.9355	46.608	14.6364	15.094	11.6711	65.998	55.5405	71.314
6 25.3	17.3224	47.323	14.9143	13.030	11.9930	62.936	55.7997	68.292
7 5.3	17.6959	48.450	15.2016	10.928	12.3862	60.095	56.1006	65.379
7 15.3	18.1140	50.081	15.5326	08.767	12.8333	57.531	56.4531	62.670
7 25.2	18.5400	51.893	15.8662	06.882	13.2945	55.594	56.8193	60.512
8 4.2	18.8908	53.709	16.1495	05.427	13.7588	54.381	57.1688	58.977
8 14.2	19.2371	55.959	16.4481	03.946	14.2565	53.448	57.5475	57.651
8 24.1	19.6014	58.482	16.7611	02.657	14.7528	53.072	57.9394	56.838
9 3.1	19.8991	60.795	17.0145	02.010	15.1937	53.655	58.2841	56.932
9 13.1	20.1421	63.196	17.2378	01.586	15.6138	54.693	58.6105	57.441
9 23.1	20.3653	65.805	17.4563	01.276	16.0159	56.095	58.9325	58.292
10 3.0	20.5713	68.311	17.6516	01.379	16.3465	58.165	59.2164	59.826
10 13.0	20.7220	70.609	17.8011	01.861	16.6041	60.752	59.4489	61.894
10 23.0	20.7909	72.747	17.8968	02.591	16.8030	63.647	59.6322	64.280
11 2.0	20.8636	74.947	17.9990	03.301	16.9469	66.583	59.7964	66.769
11 11.9	20.9240	76.899	18.0807	04.226	17.0043	69.740	59.9124	69.558
11 21.9	20.8835	78.313	18.0808	05.531	16.9639	73.148	59.9443	72.652
12 1.9	20.8138	79.666	18.0715	06.612	16.8787	76.129	59.9498	75.403
12 11.8	20.7670	80.906	18.0756	07.522	16.7377	78.759	59.9340	77.913
12 21.8	20.6589	81.516	18.0181	08.696	16.4991	81.378	59.8402	80.506
12 31.8	20.4936	81.729	17.9203	09.756	16.2090	83.493	59.6978	82.684
Mean Place	18.3024	55.392	15.4696	20.681	13.0828	82.476	56.7342	85.025
sec δ, tan δ	+1.336	+0.886	+1.006	−0.111	+1.978	−1.707	+1.367	−0.932
Dble. Trans.	October 15		October 25		November 6		November 10	

APPARENT PLACES OF STARS, 2025

TEN-DAY STARS AT UPPER TRANSIT AT GREENWICH

EQUINOX BASED RIGHT ASCENSION – WHOLE NUTATION

FK5-No.	120		153		1116		173	
HIP-No.	15863		19095		19513		22361	
Name	α Persei		174 G. Eridani		44 Tauri		Groombridge 848	
U.T.	R.A.	Dec.	R.A.	Dec.	R.A.	Dec.	R.A.	Dec.
	3^h26^m	**$+49°56'$**	**4^h06^m**	**$-27°34'$**	**4^h12^m**	**$+26°32'$**	**4^h52^m**	**$+75°58'$**
12 27.^{d}9	08.^{s}3018	69.″803	40.^{s}7084	64.″446	22.^{s}6094	50.″388	19.^{s}4190	70.″274
1 6.8	08.1974	71.209	40.6484	66.254	22.5974	50.869	19.2779	73.233
1 16.8	08.0834	72.197	40.5640	67.865	22.5709	51.131	19.0432	75.739
1 26.8	07.8704	72.556	40.4081	69.400	22.4544	51.040	18.5500	77.692
2 5.8	07.6200	72.751	40.2319	70.295	22.3067	51.073	17.9145	79.451
2 15.7	07.3852	72.716	40.0585	70.646	22.1649	51.120	17.2450	80.813
2 25.7	07.1540	72.077	39.8664	70.868	22.0095	50.776	16.5251	81.369
3 7.7	06.9023	71.042	39.6570	70.677	21.8256	50.301	15.7160	81.346
3 17.7	06.6464	69.865	39.4511	69.909	21.6310	49.921	14.8663	80.986
3 27.6	06.4878	68.534	39.2924	68.726	21.5037	49.502	14.1740	80.096
4 6.6	06.3814	66.828	39.1525	67.365	21.4015	48.836	13.5645	78.483
4 16.6	06.2594	64.930	39.0102	65.737	21.2764	48.129	12.9286	76.482
4 26.5	06.2418	63.317	38.9395	63.510	21.2338	47.752	12.5003	74.419
5 6.5	06.3445	61.718	38.9351	61.085	21.2779	47.368	12.3206	71.964
5 16.5	06.4619	59.945	38.9405	58.750	21.3231	46.825	12.2032	69.128
5 26.5	06.6328	58.535	38.9970	56.029	21.4095	46.676	12.2278	66.448
6 5.4	06.9061	57.570	39.1286	53.014	21.5805	46.738	12.4969	63.951
6 15.4	07.2465	56.680	39.3001	50.212	21.8035	46.792	12.9477	61.332
6 25.4	07.6070	55.994	39.4959	47.498	22.0416	46.952	13.4951	58.821
7 5.4	07.9772	55.796	39.7228	44.711	22.2947	47.474	14.1287	56.786
7 15.3	08.4259	56.040	40.0062	42.033	22.6167	48.219	14.9755	55.102
7 25.3	08.9071	56.406	40.3099	39.801	22.9638	48.876	15.9470	53.505
8 4.3	09.3270	56.955	40.5931	38.000	23.2659	49.557	16.8703	52.270
8 14.2	09.7701	58.069	40.9084	36.295	23.6002	50.591	17.8978	51.703
8 24.2	10.2594	59.452	41.2509	35.042	23.9767	51.634	19.0679	51.455
9 3.2	10.6931	60.761	41.5574	34.590	24.3092	52.391	20.1840	51.363
9 13.2	11.0853	62.392	41.8511	34.462	24.6202	53.297	21.2626	51.853
9 23.1	11.4757	64.404	42.1512	34.631	24.9422	54.378	22.3733	52.939
10 3.1	11.8615	66.394	42.4354	35.498	25.2642	55.214	23.4974	54.214
10 13.1	12.1932	68.361	42.6821	36.900	25.5486	55.893	24.5289	55.750
10 23.1	12.4393	70.485	42.8852	38.591	25.7762	56.648	25.4031	57.813
11 2.0	12.6955	72.811	43.0880	40.480	26.0236	57.479	26.2835	60.298
11 12.0	12.9343	74.967	43.2622	42.783	26.2595	58.065	27.1045	62.809
11 22.0	13.0486	76.873	43.3577	45.427	26.3997	58.447	27.6579	65.449
12 1.9	13.1206	78.959	43.4314	47.833	26.5193	59.072	28.0843	68.527
12 11.9	13.2042	80.965	43.4973	50.174	26.6524	59.661	28.4763	71.621
12 21.9	13.1947	82.471	43.4910	52.716	26.7070	59.894	28.6290	74.437
12 31.9	13.0947	83.776	43.4306	54.946	26.6925	60.155	28.5401	77.251
Mean Place	09.4392	59.085	40.5680	60.559	23.3496	43.331	19.0905	57.420
sec δ, tan δ	+1.554	+1.190	+1.128	−0.522	+1.118	+0.500	+4.129	+4.006
Dble. Trans.	November 12		November 22		November 23		December 3	

APPARENT PLACES OF STARS, 2025

TEN-DAY STARS AT UPPER TRANSIT AT GREENWICH

EQUINOX BASED RIGHT ASCENSION – WHOLE NUTATION

FK5-No.	201		203		210		223	
HIP-No.	25336		25769		26311		27628	
Name	γ Orionis		17 Cam		ε Orionis		β Col	
U.T.	R.A.	Dec.	R.A.	Dec.	R.A.	Dec.	R.A.	Dec.
	5^h26^m	$+6°22'$	5^h32^m	$+63°05'$	5^h37^m	$-1°10'$	5^h51^m	$-35°45'$
12 28.0 (d)	29.6865 (s)	21.051 (″)	35.0516 (s)	16.464 (″)	30.2991 (s)	69.100 (″)	52.2096 (s)	31.138 (″)
1 6.9	29.7373	20.357	35.1148	18.987	30.3536	70.253	52.2255	33.890
1 16.9	29.7652	19.583	35.1224	21.172	30.3837	71.454	52.2022	36.591
1 26.9	29.7042	18.733	34.9582	22.987	30.3269	72.673	52.0981	39.156
2 5.8	29.6069	18.260	34.7106	24.764	30.2331	73.454	51.9488	41.096
2 15.8	29.5029	18.004	34.4388	26.263	30.1303	73.954	51.7784	42.534
2 25.8	29.3722	17.567	34.1150	27.078	29.9996	74.566	51.5756	43.847
3 7.8	29.2033	17.273	33.7168	27.457	29.8303	74.949	51.3363	44.650
3 17.7	29.0147	17.329	33.2787	27.617	29.6406	74.899	51.0791	44.741
3 27.7	28.8727	17.441	32.9244	27.275	29.4933	74.733	50.8494	44.470
4 6.7	28.7393	17.458	32.5921	26.265	29.3526	74.589	50.6263	43.945
4 16.7	28.5778	17.650	32.2159	24.946	29.1848	74.192	50.3915	42.910
4 26.6	28.4830	18.212	31.9627	23.541	29.0799	73.380	50.2094	41.255
5 6.6	28.4564	18.754	31.8436	21.692	29.0393	72.547	50.0845	39.386
5 16.6	28.4258	19.203	31.7329	19.461	28.9956	71.760	49.9726	37.397
5 26.5	28.4352	20.016	31.7025	17.344	28.9906	70.579	49.9041	34.884
6 5.5	28.5193	21.090	31.8181	15.311	29.0579	69.123	49.9056	32.013
6 15.5	28.6459	22.008	32.0289	13.062	29.1674	67.827	49.9558	29.267
6 25.5	28.7905	22.988	32.2862	10.850	29.2955	66.472	50.0385	26.444
7 5.4	28.9573	24.262	32.5946	09.032	29.4468	64.829	50.1590	23.359
7 15.4	29.1925	25.554	33.0382	07.421	29.6650	63.203	50.3451	20.378
7 25.4	29.4556	26.592	33.5493	05.780	29.9121	61.863	50.5698	17.778
8 4.4	29.6899	27.577	34.0230	04.436	30.1345	60.607	50.7960	15.389
8 14.3	29.9649	28.715	34.5765	03.635	30.3980	59.256	51.0660	13.085
8 24.3	30.2859	29.610	35.2243	03.006	30.7074	58.213	51.3817	11.292
9 3.3	30.5780	30.060	35.8331	02.453	30.9915	57.672	51.6910	10.222
9 13.2	30.8637	30.498	36.4336	02.402	31.2717	57.208	52.0049	09.465
9 23.2	31.1708	30.894	37.0743	02.841	31.5742	56.862	52.3401	09.084
10 3.2	31.4864	30.804	37.7343	03.368	31.8859	57.081	52.6824	09.546
10 13.2	31.7773	30.410	38.3493	04.111	32.1753	57.671	53.0060	10.628
10 23.1	32.0288	30.002	38.8876	05.367	32.4284	58.335	53.2995	12.027
11 2.1	32.3054	29.468	39.4608	06.982	32.7054	59.202	53.5993	13.887
11 12.1	32.5747	28.555	40.0135	08.600	32.9752	60.505	53.8805	16.390
11 22.1	32.7644	27.472	40.4125	10.420	33.1691	62.010	54.0933	19.259
12 2.0	32.9405	26.613	40.7666	12.718	33.3493	63.327	54.2772	22.086
12 12.0	33.1269	25.669	41.1228	15.056	33.5375	64.759	54.4459	25.124
12 22.0	33.2407	24.482	41.3352	17.235	33.6552	66.432	54.5433	28.443
12 31.9	33.2910	23.503	41.4127	19.555	33.7105	67.877	54.5736	31.509
Mean Place	30.0225	14.575	35.1625	0 5.524	30.5263	75.443	51.6051	36.449
sec δ, tan δ	+1.006	+0.112	+2.209	+1.970	+1.000	−0.021	+1.232	−0.720
Dble. Trans.	December 12		December 14		December 15		December 19	

APPARENT PLACES OF STARS, 2025

TEN-DAY STARS AT UPPER TRANSIT AT GREENWICH

EQUINOX BASED RIGHT ASCENSION – WHOLE NUTATION

FK5-No.	1166		243		245		254	
HIP-No.	29134		30324		30438		32246	
Name	ν Doradus		β Canis Majoris		α Carinae		ϵ Geminorum	
U.T.	R.A.	Dec.	R.A.	Dec.	R.A.	Dec.	R.A.	Dec.
	6^h08^m	**$-68°50'$**	**6^h23^m**	**$-17°57'$**	**6^h24^m**	**$-52°42'$**	**6^h45^m**	**$+25°06'$**
12 28.0	38.4356	47.972	49.4704	64.546	32.6201	28.849	29.6659	21.359
1 7.0	38.3018	51.337	49.5499	66.758	32.6284	32.196	29.8120	21.583
1 16.9	38.0539	54.661	49.5989	69.004	32.5761	35.545	29.9289	21.690
1 26.9	37.7085	57.820	49.5631	71.154	32.4385	38.742	29.9395	21.802
2 5.9	37.2611	60.326	49.4829	72.784	32.2338	41.325	29.9021	22.224
2 15.9	36.7345	62.289	49.3839	74.042	31.9857	43.406	29.8472	22.720
2 25.8	36.1470	64.069	49.2519	75.286	31.6954	45.331	29.7498	22.929
3 7.8	35.5134	65.259	49.0762	76.132	31.3604	46.688	29.5950	23.166
3 17.8	34.8607	65.646	48.8737	76.375	30.9992	47.258	29.4041	23.610
3 27.7	34.1958	65.607	48.7012	76.409	30.6483	47.438	29.2520	23.843
4 6.7	33.5459	65.221	48.5292	76.306	30.3020	47.285	29.0933	23.779
4 16.7	32.9381	64.214	48.3294	75.752	29.9535	46.499	28.8885	23.789
4 26.7	32.3689	62.520	48.1806	74.704	29.6448	45.049	28.7478	23.889
5 6.6	31.8630	60.541	48.0871	73.549	29.3900	43.323	28.6712	23.703
5 16.6	31.4418	58.348	47.9927	72.293	29.1663	41.359	28.5794	23.341
5 26.6	31.1007	55.565	47.9329	70.560	28.9897	38.797	28.5254	23.180
6 5.6	30.8547	52.375	47.9390	68.516	28.8841	35.823	28.5500	23.080
6 15.5	30.7094	49.265	47.9881	66.595	28.8417	32.908	28.6195	22.720
6 25.5	30.6662	46.032	48.0592	64.565	28.8500	29.836	28.7082	22.383
7 5.5	30.7324	42.502	48.1565	62.218	28.9133	26.431	28.8066	22.364
7 15.4	30.8930	39.083	48.3201	59.945	29.0500	23.118	29.0014	22.186
7 25.4	31.1518	36.041	48.5181	57.981	29.2455	20.149	29.2270	21.829
8 4.4	31.5114	33.200	48.7043	56.090	29.4739	17.321	29.4249	21.591
8 14.4	31.9435	30.482	48.9346	54.203	29.7550	14.593	29.6777	21.552
8 24.3	32.4458	28.323	49.2147	52.741	30.0934	12.400	29.9908	21.319
9 3.3	33.0128	26.924	49.4837	51.846	30.4542	10.916	30.2810	20.869
9 13.3	33.6164	25.898	49.7572	51.138	30.8332	09.776	30.5763	20.581
9 23.2	34.2464	25.326	50.0577	50.700	31.2393	09.070	30.9088	20.383
10 3.2	34.8821	25.688	50.3750	50.992	31.6615	09.284	31.2644	19.873
10 13.2	35.5043	26.751	50.6786	51.785	32.0743	10.181	31.6048	19.271
10 23.2	36.1019	28.212	50.9549	52.771	32.4642	11.456	31.9148	18.901
11 2.1	36.6399	30.251	51.2548	54.165	32.8470	13.327	32.2679	18.512
11 12.1	37.1088	33.030	51.5502	56.149	33.2045	15.951	32.6238	17.881
11 22.1	37.5029	36.241	51.7793	58.401	33.4985	18.998	32.9004	17.354
12 2.1	37.7894	39.499	51.9926	60.598	33.7422	22.123	33.1730	17.154
12 12.0	37.9650	43.053	52.2084	63.038	33.9463	25.585	33.4640	16.890
12 22.0	38.0310	46.929	52.3579	65.739	34.0731	29.384	33.6774	16.552
12 32.0	37.9744	50.582	52.4444	68.212	34.1166	32.986	33.8216	16.520
Mean Place	34.4517	55.065	49.4061	73.082	31.1010	37.732	30.0071	12.658
sec δ, tan δ	+2.771	–2.585	+1.051	–0.324	+1.651	–1.313	+1.104	+0.469
Dble. Trans.	December 23		December 27		December 27		January 1	

(First data row units: $28^d.0$, $38^s.4356$, $47''.972$, etc.)

APPARENT PLACES OF STARS, 2025

TEN-DAY STARS AT UPPER TRANSIT AT GREENWICH

EQUINOX BASED RIGHT ASCENSION – WHOLE NUTATION

FK5-No.	306		323		348		354	
HIP-No.	39429		42452		45238		46390	
Name	ζ Puppis		Groombridge 1460		β Carinae		α Hydrae	
U.T.	R.A.	Dec.	R.A.	Dec.	R.A.	Dec.	R.A.	Dec.
	8h04m	**– 40°04′**	**8h41m**	**+ 52°36′**	**9h13m**	**– 69°48′**	**9h28m**	**– 8°45′**
12 28.1d	29.1632s	16.706″	10.2324s	75.569″	29.6679s	54.901″	49.8019s	58.434″
1 7.0	29.3330	20.052	10.5951	76.745	30.0334	58.405	50.0625	60.721
1 17.0	29.4637	23.601	10.9081	77.988	30.3153	62.351	50.3045	63.145
1 27.0	29.5098	27.035	11.0775	79.559	30.4939	66.322	50.4626	65.318
2 6.0	29.4879	30.072	11.1829	81.508	30.5136	70.172	50.5706	67.164
2 15.9	29.4234	32.824	11.2471	83.510	30.4088	73.963	50.6505	68.840
2 25.9	29.3174	35.518	11.2230	85.304	30.2352	77.800	50.6886	70.476
3 7.9	29.1506	37.733	11.0996	87.128	29.9481	81.267	50.6616	71.729
3 17.8	28.9327	39.251	10.9107	89.037	29.5387	84.134	50.5796	72.476
3 27.8	28.7214	40.561	10.7334	90.488	29.0752	86.892	50.5091	73.242
4 6.8	28.4981	41.573	10.5041	91.480	28.5801	89.327	50.4131	73.869
4 16.8	28.2355	41.874	10.1935	92.411	28.0250	90.965	50.2545	73.945
4 26.7	27.9947	41.657	09.9417	93.013	27.4193	92.135	50.1197	73.842
5 6.7	27.7936	41.217	09.7331	92.962	26.8323	93.008	50.0150	73.774
5 16.7	27.5925	40.377	09.4797	92.568	26.2720	93.275	49.8807	73.425
5 26.7	27.4088	38.929	09.2658	92.012	25.6958	92.853	49.7524	72.739
6 5.6	27.2746	37.073	09.1394	91.087	25.1428	91.921	49.6653	71.935
6 15.6	27.1876	35.150	09.0515	89.656	24.6748	90.705	49.6053	71.183
6 25.6	27.1256	32.908	08.9806	88.018	24.2618	88.960	49.5475	70.231
7 5.5	27.0878	30.176	08.9531	86.411	23.8843	86.532	49.4953	68.939
7 15.5	27.1174	27.469	09.0301	84.474	23.5956	83.975	49.5042	67.801
7 25.5	27.1975	24.913	09.1474	82.184	23.4267	81.321	49.5435	66.809
8 4.5	27.2866	22.191	09.2497	80.030	23.3425	78.217	49.5635	65.577
8 14.4	27.4248	19.483	09.4466	77.918	23.3290	75.011	49.6307	64.400
8 24.4	27.6291	17.189	09.7296	75.538	23.4429	72.047	49.7584	63.579
9 3.4	27.8568	15.312	09.9990	73.138	23.6917	69.233	49.8884	62.931
9 13.4	28.1062	13.637	10.3063	70.985	24.0117	66.512	50.0372	62.359
9 23.3	28.3962	12.325	10.6898	68.968	24.4109	64.086	50.2322	62.042
10 3.3	28.7313	11.828	11.1183	66.845	24.9373	62.375	50.4752	62.292
10 13.3	29.0780	11.905	11.5504	64.898	25.5427	61.182	50.7317	62.855
10 23.2	29.4151	12.275	11.9819	63.455	26.1749	60.279	50.9847	63.483
11 2.2	29.7848	13.353	12.4940	62.157	26.8484	60.212	51.2982	64.651
11 12.2	30.1679	15.231	13.0215	60.903	27.5740	61.030	51.6427	66.367
11 22.2	30.5053	17.466	13.4763	60.197	28.2907	62.263	51.9478	68.126
12 2.1	30.8220	19.973	13.9576	60.023	28.9408	64.037	52.2657	70.041
12 12.1	31.1387	23.043	14.4703	59.969	29.5602	66.626	52.6166	72.398
12 22.1	31.4009	26.490	14.8922	60.263	30.1354	69.738	52.9248	74.869
12 32.1	31.5934	29.893	15.2424	61.146	30.5896	73.069	53.1844	77.180
Mean Place	28.8588	34.199	10.1491	72.170	28.1490	80.974	50.4231	73.128
sec δ, tan δ	+1.307	–0.841	+1.647	+1.309	+2.899	–2.721	+1.012	–0.154
Dble. Trans.	January 21		January 30		February 7		February 11	

APPARENT PLACES OF STARS, 2025

TEN-DAY STARS AT UPPER TRANSIT AT GREENWICH

EQUINOX BASED RIGHT ASCENSION – WHOLE NUTATION

FK5-No.	363		1260		1275		422	
HIP-No.	47594		49339		52098		54872	
Name	Groombridge 1564		193 G. Hydrae		37 LMi		δ Leonis	
U.T.	R.A.	Dec.	R.A.	Dec.	R.A.	Dec.	R.A.	Dec.
	9h44m	**+ 69°06′**	**10h05m**	**– 24°24′**	**10h40m**	**+ 31°50′**	**11h15m**	**+ 20°22′**
12 28.d1	24.s3819	68.″743	30.s9288	16.″016	08.s2468	36.″832	26.s6736	64.″973
1 7.1	25.0510	69.997	31.2183	18.825	08.6136	35.919	27.0258	63.379
1 17.1	25.6329	71.480	31.4926	21.894	08.9627	35.196	27.3701	61.927
1 27.1	26.0187	73.509	31.6888	24.787	09.2190	35.058	27.6349	61.041
2 6.0	26.3199	76.001	31.8293	27.486	09.4363	35.388	27.8651	60.562
2 16.0	26.5436	78.610	31.9368	30.096	09.6275	35.952	28.0739	60.309
2 26.0	26.6048	81.155	32.0059	32.668	09.7575	36.687	28.2331	60.292
3 7.9	26.5064	83.814	32.0073	34.860	09.8094	37.801	28.3215	60.713
3 17.9	26.3054	86.556	31.9455	36.538	09.8014	39.261	28.3515	61.530
3 27.9	26.0670	88.814	31.8896	38.208	09.7940	40.530	28.3842	62.241
4 6.9	25.7046	90.632	31.8070	39.637	09.7345	41.766	28.3705	63.062
4 16.8	25.2161	92.364	31.6567	40.406	09.5940	43.283	28.2771	64.280
4 26.8	24.7768	93.605	31.5174	40.945	09.4779	44.574	28.2011	65.346
5 6.8	24.3426	94.078	31.4027	41.392	09.3740	45.464	28.1349	66.153
5 16.8	23.8219	94.158	31.2574	41.386	09.2145	46.327	28.0138	67.085
5 26.7	23.3480	93.891	31.1071	40.948	09.0603	47.077	27.8890	67.981
6 5.7	22.9759	93.037	30.9867	40.280	08.9465	47.460	27.7936	68.595
6 15.7	22.6301	91.561	30.8926	39.505	08.8426	47.463	27.7043	68.952
6 25.6	22.3054	89.747	30.7975	38.388	08.7303	47.320	27.6007	69.258
7 5.6	22.0626	87.772	30.6994	36.818	08.6269	47.107	27.4944	69.540
7 15.6	21.9590	85.285	30.6580	35.310	08.5872	46.419	27.4416	69.387
7 25.6	21.9032	82.357	30.6501	33.807	08.5647	45.381	27.4023	68.966
8 4.5	21.8612	79.495	30.6249	31.940	08.5169	44.413	27.3326	68.652
8 14.5	21.9765	76.513	30.6425	30.098	08.5311	43.184	27.3137	68.046
8 24.5	22.2114	73.174	30.7238	28.549	08.6052	41.500	27.3493	67.006
9 3.5	22.4486	69.847	30.8179	27.085	08.6697	39.717	27.3755	65.899
9 13.4	22.7773	66.719	30.9328	25.694	08.7646	37.889	27.4254	64.671
9 23.4	23.2416	63.667	31.0954	24.580	08.9215	35.857	27.5311	63.154
10 3.4	23.7765	60.564	31.3173	24.041	09.1268	33.511	27.6871	61.279
10 13.3	24.3416	57.717	31.5629	23.854	09.3497	31.141	27.8629	59.311
10 23.3	24.9570	55.423	31.8094	23.817	09.5863	28.966	28.0511	57.417
11 2.3	25.7016	53.311	32.1211	24.463	09.9061	26.564	28.3211	55.156
11 12.3	26.4710	51.378	32.4747	25.758	10.2626	24.039	28.6337	52.680
11 22.2	27.1821	50.170	32.7987	27.210	10.5878	21.924	28.9222	50.499
12 2.2	27.9650	49.561	33.1349	29.030	10.9575	20.001	29.2549	48.327
12 12.2	28.7894	49.182	33.5081	31.466	11.3770	18.020	29.6418	45.974
12 22.2	29.5016	49.379	33.8487	34.142	11.7548	16.419	29.9990	43.916
12 32.1	30.1417	50.315	34.1417	36.836	12.1005	15.346	30.3306	42.259
Mean Place	23.2423	70.502	31.7204	35.335	08.9349	34.833	27.6685	60.914
sec δ, tan δ	+2.806	+2.621	+1.098	–0.454	+1.177	+0.621	+1.067	+0.372
Dble. Trans.	February 15		February 21		March 1		March 10	

APPARENT PLACES OF STARS, 2025

TEN-DAY STARS AT UPPER TRANSIT AT GREENWICH

EQUINOX BASED RIGHT ASCENSION – WHOLE NUTATION

FK5-No.	423		426		1307		455	
HIP-No.	54879		55282		57939		59747	
Name	ϑ Leonis		δ Crateris		Groombridge 1830		δ Crucis	
U.T.	R.A.	Dec.	R.A.	Dec.	R.A.	Dec.	R.A.	Dec.
	$11^{\mathrm{h}}15^{\mathrm{m}}$	**$+15°17'$**	**$11^{\mathrm{h}}20^{\mathrm{m}}$**	**$-14°54'$**	**$11^{\mathrm{h}}54^{\mathrm{m}}$**	**$+37°31'$**	**$12^{\mathrm{h}}16^{\mathrm{m}}$**	**$-58°52'$**
12 $28^{\mathrm{d}}.2$	$33^{\mathrm{s}}.5007$	$28''.928$	$35^{\mathrm{s}}.7741$	$45''.001$	$25^{\mathrm{s}}.4004$	$68''.211$	$28^{\mathrm{s}}.4022$	$59''.189$
1 7.2	33.8439	27.177	36.1045	47.460	25.8207	66.653	28.9322	61.204
1 17.1	34.1807	25.522	36.4335	50.111	26.2336	65.415	29.4752	63.806
1 27.1	34.4400	24.394	36.6922	52.491	26.5677	64.914	29.9530	66.513
2 6.1	34.6640	23.641	36.9073	54.738	26.8748	64.918	30.3505	69.578
2 16.1	34.8670	23.092	37.0982	56.944	27.1607	65.230	30.6963	72.995
2 26.0	35.0235	22.766	37.2543	59.042	27.3862	65.913	31.0152	76.572
3 8.0	35.1109	22.874	37.3442	60.777	27.5351	67.100	31.2412	80.086
3 18.0	35.1401	23.387	37.3699	62.100	27.6239	68.673	31.3510	83.448
3 28.0	35.1734	23.808	37.4023	63.473	27.7037	70.155	31.4492	87.011
4 6.9	35.1633	24.364	37.4017	64.602	27.7202	71.761	31.5059	90.378
4 16.9	35.0746	25.350	37.3230	65.139	27.6489	73.701	31.4433	93.173
4 26.9	35.0025	26.219	37.2518	65.619	27.5894	75.377	31.3384	95.942
5 6.8	34.9416	26.872	37.1960	66.074	27.5233	76.711	31.2337	98.554
5 16.8	34.8278	27.699	37.0949	66.094	27.3885	78.082	31.0707	100.516
5 26.8	34.7089	28.536	36.9791	65.845	27.2466	79.249	30.8429	102.079
6 5.8	34.6182	29.143	36.8828	65.526	27.1287	79.953	30.5944	103.350
6 15.7	34.5347	29.542	36.7995	65.123	27.0041	80.281	30.3610	104.195
6 25.7	34.4368	29.937	36.7013	64.439	26.8586	80.410	30.0962	104.452
7 5.7	34.3339	30.355	36.5871	63.439	26.7128	80.310	29.7744	104.092
7 15.7	34.2834	30.379	36.5209	62.578	26.6175	79.623	29.4947	103.532
7 25.6	34.2470	30.173	36.4756	61.697	26.5269	78.564	29.2559	102.540
8 4.6	34.1795	30.107	36.3988	60.448	26.4070	77.479	28.9837	100.800
8 14.6	34.1601	29.770	36.3596	59.307	26.3446	75.956	28.7384	98.883
8 24.5	34.1954	29.020	36.3776	58.430	26.3341	73.914	28.5791	96.861
9 3.5	34.2223	28.228	36.3989	57.507	26.3124	71.775	28.4697	94.403
9 13.5	34.2693	27.307	36.4379	56.645	26.3236	69.470	28.3910	91.783
9 23.5	34.3714	26.080	36.5243	56.042	26.3984	66.832	28.3768	89.223
10 3.4	34.5246	24.486	36.6719	55.855	26.5234	63.883	28.4868	86.847
10 13.4	34.6977	22.776	36.8459	55.895	26.6731	60.899	28.6708	84.562
10 23.4	34.8812	21.113	37.0258	56.032	26.8494	58.016	28.8784	82.344
11 2.4	35.1452	19.039	37.2835	56.792	27.1183	54.845	29.2013	80.793
11 12.3	35.4527	16.708	37.5937	58.044	27.4330	51.608	29.6376	79.782
11 22.3	35.7360	14.622	37.8829	59.320	27.7348	48.800	30.0854	78.936
12 2.3	36.0610	12.489	38.2026	60.970	28.0992	46.103	30.5658	78.792
12 12.2	36.4400	10.121	38.5764	63.170	28.5252	43.383	31.1302	79.485
12 22.2	36.7907	07.992	38.9301	65.460	28.9268	41.156	31.7104	80.563
12 32.2	37.1150	06.209	39.2518	67.747	29.3160	39.478	32.2515	82.092
Mean Place	34.5451	23.240	37.0786	60.770	26.4993	68.007	30.8995	86.254
sec δ, tan δ	+1.037	+0.273	+1.035	−0.266	+1.261	+0.768	+1.935	−1.657
Dble. Trans.	March 10		March 12		March 20		March 26	

APPARENT PLACES OF STARS, 2025

TEN-DAY STARS AT UPPER TRANSIT AT GREENWICH

EQUINOX BASED RIGHT ASCENSION – WHOLE NUTATION

FK5-No.	464		468		479		1357	
HIP-No.	60823		61084		62131		67057	
Name	σ Centauri		γ Crucis		330 G. Hydrae		83 Vir	
U.T.	R.A.	Dec.	R.A.	Dec.	R.A.	Dec.	R.A.	Dec.
	$12^{h}29^{m}$	**$-50°21'$**	**$12^{h}32^{m}$**	**$-57°14'$**	**$12^{h}45^{m}$**	**$-28°27'$**	**$13^{h}45^{m}$**	**$-16°18'$**
12 $28^{d}.3$	$23^{s}.4372$	$53''.029$	$33^{s}.1025$	$54''.237$	$20^{s}.3619$	$31''.085$	$50^{s}.3390$	$12''.918$
1 7.2	23.8981	55.035	33.6244	56.105	20.7370	33.229	50.6923	14.899
1 17.2	24.3759	57.562	34.1652	58.545	21.1294	35.683	51.0732	17.020
1 27.2	24.7926	60.137	34.6442	61.086	21.4647	37.988	51.4067	18.867
2 6.1	25.1481	63.029	35.0510	64.006	21.7620	40.430	51.7191	20.801
2 16.1	25.4678	66.235	35.4134	67.297	22.0412	43.025	52.0268	22.818
2 26.1	25.7644	69.549	35.7523	70.741	22.2979	45.558	52.3182	24.629
3 8.1	25.9812	72.767	36.0025	74.141	22.4903	47.861	52.5547	26.157
3 18.0	26.1016	75.827	36.1412	77.426	22.6114	49.934	52.7306	27.491
3 28.0	26.2210	79.070	36.2730	80.927	22.7415	52.113	52.9210	28.875
4 7.0	26.3040	82.100	36.3643	84.240	22.8386	54.024	53.0790	29.936
4 17.0	26.2795	84.573	36.3362	87.012	22.8448	55.393	53.1486	30.530
4 26.9	26.2298	87.046	36.2697	89.796	22.8469	56.798	53.2189	31.240
5 6.9	26.1851	89.377	36.2038	92.437	22.8576	58.110	53.2946	31.875
5 16.9	26.0821	91.085	36.0755	94.443	22.8113	58.884	53.3071	32.037
5 26.8	25.9273	92.443	35.8827	96.088	22.7315	59.428	53.2861	32.121
6 5.8	25.7607	93.563	35.6686	97.473	22.6529	59.883	53.2631	32.263
6 15.8	25.6039	94.296	35.4652	98.436	22.5774	60.089	53.2330	32.224
6 25.8	25.4149	94.489	35.2257	98.827	22.4712	59.912	53.1644	31.928
7 5.7	25.1744	94.137	34.9255	98.633	22.3254	59.392	53.0506	31.501
7 15.7	24.9726	93.637	34.6640	98.246	22.2154	58.899	52.9641	31.216
7 25.7	24.7988	92.742	34.4368	97.417	22.1184	58.170	52.8769	30.758
8 4.7	24.5831	91.155	34.1680	95.843	21.9735	56.936	52.7307	29.946
8 14.6	24.3918	89.446	33.9231	94.102	21.8534	55.753	52.6027	29.327
8 24.6	24.2720	87.663	33.7587	92.237	21.7870	54.640	52.5159	28.820
9 3.6	24.1800	85.466	33.6351	89.904	21.7225	53.239	52.4161	28.044
9 13.5	24.1095	83.138	33.5380	87.400	21.6710	51.822	52.3221	27.329
9 23.5	24.0939	80.895	33.5019	84.941	21.6642	50.596	52.2664	26.859
10 3.5	24.1796	78.830	33.5850	82.622	21.7286	49.593	52.2711	26.540
10 13.5	24.3219	76.846	33.7380	80.357	21.8274	48.683	52.3034	26.274
10 23.4	24.4796	74.932	33.9122	78.142	21.9344	47.858	52.3371	26.159
11 2.4	24.7430	73.665	34.2032	76.566	22.1336	47.642	52.4660	26.432
11 12.4	25.1032	72.893	34.6070	75.482	22.4059	47.837	52.6670	27.031
11 22.3	25.4658	72.253	35.0214	74.537	22.6694	48.078	52.8601	27.601
12 2.3	25.8655	72.284	35.4742	74.286	22.9754	48.874	53.1062	28.666
12 12.3	26.3467	73.101	36.0162	74.848	23.3560	50.297	53.4337	30.203
12 22.3	26.8375	74.240	36.5759	75.766	23.7363	51.864	53.7667	31.722
12 32.2	27.2981	75.781	37.1026	77.136	24.0971	53.648	54.0943	33.371
Mean Place	25.8463	77.660	35.7913	80.425	22.4628	48.637	52.6650	22.977
sec δ, tan δ	+1.568	−1.208	+1.849	−1.555	+1.138	−0.542	+1.042	−0.293
Dble. Trans.	March 29		March 30		April 2		April 18	

APPARENT PLACES OF STARS, 2025

TEN-DAY STARS AT UPPER TRANSIT AT GREENWICH

EQUINOX BASED RIGHT ASCENSION – WHOLE NUTATION

FK5-No.	509		529		1396		564	
HIP-No.	67301		70069		73996		74785	
Name	η Ursae Majoris		ν Cen		45 Bootis		β Librae	
U.T.	R.A.	Dec.	R.A.	Dec.	R.A.	Dec.	R.A.	Dec.
	$13^{h}48^{m}$	**$+49°10'$**	**$14^{h}22^{m}$**	**$-56°29'$**	**$15^{h}08^{m}$**	**$+24°45'$**	**$15^{h}18^{m}$**	**$-9°28'$**
12 $28^{d}.3$	$31^{s}.1938$	$62''.056$	$03^{s}.5046$	$50''.476$	$23^{s}.1336$	$69''.519$	$20^{s}.2630$	$29''.649$
1 7.3	31.6389	59.694	04.0296	51.004	23.4497	66.717	20.5741	31.436
1 17.3	32.0992	57.820	04.6129	52.014	23.8015	64.239	20.9280	33.210
1 27.2	32.5157	56.818	05.1548	53.130	24.1268	62.407	21.2500	34.660
2 6.2	32.9339	56.340	05.6629	54.795	24.4585	60.844	21.5703	36.207
2 16.2	33.3457	56.298	06.1622	56.962	24.8010	59.582	21.9024	37.788
2 26.1	33.7066	56.927	06.6611	59.279	25.1299	58.974	22.2318	39.025
3 8.1	34.0039	58.249	07.0890	61.719	25.4194	58.989	22.5208	39.957
3 18.1	34.2491	60.030	07.4189	64.337	25.6683	59.389	22.7620	40.769
3 28.1	34.4716	61.967	07.7657	67.270	25.9261	60.059	23.0272	41.526
4 7.0	34.6174	64.334	08.0759	70.108	26.1449	61.321	23.2664	41.871
4 17.0	34.6720	67.131	08.2544	72.679	26.2865	63.072	23.4219	41.852
4 27.0	34.7159	69.728	08.4057	75.512	26.4276	64.754	23.5825	41.980
5 7.0	34.7204	72.199	08.5554	78.303	26.5565	66.588	23.7473	41.972
5 16.9	34.6367	74.848	08.6163	80.599	26.6153	68.806	23.8449	41.537
5 26.9	34.5239	77.243	08.6005	82.804	26.6417	70.914	23.9068	41.152
6 5.9	34.4022	79.141	08.5498	84.957	26.6567	72.742	23.9601	40.916
6 15.8	34.2420	80.753	08.4841	86.720	26.6446	74.564	23.9962	40.503
6 25.8	34.0377	82.134	08.3512	88.032	26.5857	76.353	23.9841	39.935
7 5.8	33.8138	83.080	08.1236	88.988	26.4832	77.822	23.9157	39.450
7 15.8	33.6094	83.350	07.9141	89.777	26.3905	78.829	23.8622	39.159
7 25.7	33.3831	83.257	07.7043	90.051	26.2756	79.716	23.7928	38.707
8 4.7	33.1190	82.954	07.4049	89.634	26.1001	80.479	23.6493	38.079
8 14.7	32.8961	81.971	07.1077	89.079	25.9366	80.637	23.5114	37.766
8 24.7	32.7023	80.400	06.8663	88.242	25.7917	80.386	23.3987	37.552
9 3.6	32.4913	78.667	06.6246	86.735	25.6189	80.076	23.2567	37.091
9 13.6	32.3129	76.544	06.3862	84.973	25.4497	79.359	23.1086	36.779
9 23.6	32.1937	73.874	06.1937	83.131	25.3113	78.090	22.9868	36.761
10 3.5	32.1206	70.868	06.1071	81.085	25.2123	76.543	22.9129	36.783
10 13.5	32.0804	67.744	06.0785	78.828	25.1327	74.835	22.8579	36.798
10 23.5	32.0865	64.479	06.0618	76.501	25.0678	72.806	22.8037	36.995
11 2.5	32.1939	60.806	06.1799	74.546	25.0875	70.256	22.8389	37.612
11 12.4	32.3589	57.103	06.4242	72.732	25.1662	67.604	22.9401	38.282
11 22.4	32.5455	53.720	06.6816	70.879	25.2511	65.017	23.0358	38.976
12 2.4	32.8239	50.266	07.0114	69.633	25.4058	62.062	23.1954	40.136
12 12.4	33.1819	46.815	07.4694	68.979	25.6423	58.941	23.4446	41.597
12 22.3	33.5504	43.920	07.9666	68.497	25.8958	56.138	23.7092	42.932
12 32.3	33.9462	41.514	08.4647	68.473	26.1705	53.492	23.9839	44.386
Mean Place	32.5769	72.074	07.6188	68.653	25.3123	76.153	22.9612	31.440
sec δ, tan δ	+1.530	+1.158	+1.812	−1.511	+1.101	+0.461	+1.014	−0.167
Dble. Trans.	April 18		April 27		May 9		May 11	

APPARENT PLACES OF STARS, 2025

TEN-DAY STARS AT UPPER TRANSIT AT GREENWICH

EQUINOX BASED RIGHT ASCENSION – WHOLE NUTATION

FK5-No.	560		588		603		622	
HIP-No.	74946		77622		79593		81377	
Name	γ Trianguli A		ϵ Ser		δ Ophiuchi		ζ Ophiuchi	
U.T.	R.A.	Dec.	R.A.	Dec.	R.A.	Dec.	R.A.	Dec.
	$15^{\mathrm{h}}21^{\mathrm{m}}$	**$-68°46'$**	**$15^{\mathrm{h}}52^{\mathrm{m}}$**	**$+4°23'$**	**$16^{\mathrm{h}}15^{\mathrm{m}}$**	**$-3°45'$**	**$16^{\mathrm{h}}38^{\mathrm{m}}$**	**$-10°37'$**
12 $28^{\mathrm{d}}.4$	$13^{\mathrm{s}}.2120$	$01''.681$	$02^{\mathrm{s}}.7440$	$65''.801$	$38^{\mathrm{s}}.2525$	$32''.510$	$30^{\mathrm{s}}.9319$	$03''.748$
1 7.3	13.9068	01.025	03.0245	63.569	38.5207	34.341	31.1897	05.147
1 17.3	14.7117	00.823	03.3501	61.501	38.8389	36.058	31.5020	06.461
1 27.3	15.4920	00.789	03.6516	59.864	39.1347	37.434	31.7938	07.507
2 6.3	16.2563	01.431	03.9602	58.253	39.4397	38.884	32.0972	08.705
2 16.2	17.0329	02.693	04.2864	56.753	39.7663	40.307	32.4263	09.943
2 26.2	17.8324	04.175	04.6131	55.763	40.0989	41.288	32.7666	10.797
3 8.2	18.5561	05.951	04.9068	55.207	40.4017	41.931	33.0797	11.401
3 18.2	19.1591	08.134	05.1615	54.858	40.6665	42.465	33.3565	11.987
3 28.1	19.7935	10.736	05.4402	54.713	40.9618	42.835	33.6694	12.448
4 7.1	20.3835	13.373	05.6941	55.107	41.2372	42.715	33.9666	12.471
4 17.1	20.7934	15.974	05.8716	55.894	41.4357	42.274	34.1865	12.252
4 27.0	21.1601	18.998	06.0551	56.598	41.6432	41.944	34.4184	12.178
5 7.0	21.5132	22.078	06.2397	57.510	41.8562	41.404	34.6595	11.905
5 17.0	21.7319	24.801	06.3583	58.844	42.0034	40.458	34.8344	11.259
5 27.0	21.8298	27.599	06.4430	60.100	42.1159	39.603	34.9740	10.728
6 5.9	21.8563	30.463	06.5173	61.172	42.2177	38.914	35.1028	10.362
6 15.9	21.8396	32.967	06.5699	62.383	42.2986	38.044	35.2108	09.794
6 25.9	21.7082	35.100	06.5732	63.670	42.3279	37.075	35.2648	09.121
7 5.9	21.4197	36.992	06.5215	64.733	42.2969	36.311	35.2539	08.653
7 15.8	21.1285	38.695	06.4787	65.515	42.2736	35.760	35.2495	08.346
7 25.8	20.8108	39.831	06.4139	66.364	42.2269	35.072	35.2196	07.854
8 4.8	20.3452	40.295	06.2752	67.211	42.0995	34.348	35.1025	07.307
8 14.7	19.8576	40.578	06.1384	67.604	41.9696	34.013	34.9791	07.097
8 24.7	19.4216	40.443	06.0186	67.793	41.8551	33.790	34.8691	06.920
9 3.7	18.9657	39.499	05.8644	68.089	41.7019	33.387	34.7159	06.505
9 13.7	18.4985	38.198	05.7016	68.093	41.5350	33.216	34.5441	06.270
9 23.6	18.0779	36.678	05.5601	67.678	41.3854	33.393	34.3860	06.315
10 3.6	17.7923	34.703	05.4578	67.149	41.2742	33.590	34.2650	06.290
10 13.6	17.5828	32.324	05.3696	66.550	41.1747	33.796	34.1531	06.213
10 23.6	17.3888	29.751	05.2823	65.655	41.0708	34.267	34.0322	06.366
11 2.5	17.3830	27.322	05.2778	64.308	41.0501	35.126	33.9950	06.833
11 12.5	17.5601	24.794	05.3353	62.905	41.0932	35.985	34.0230	07.239
11 22.5	17.7708	22.107	05.3900	61.497	41.1302	36.846	34.0427	07.631
12 2.4	18.1004	19.919	05.5080	59.622	41.2293	38.188	34.1236	08.471
12 12.4	18.6264	18.161	05.7132	57.524	41.4194	39.750	34.2985	09.532
12 22.4	19.2339	16.476	05.9365	55.614	41.6292	41.155	34.4952	10.446
12 32.4	19.8747	15.255	06.1772	53.657	41.8566	42.674	34.7106	11.513
Mean Place	19.7474	15.483	05.3593	69.554	41.0710	29.250	33.9421	0 0.139
sec δ, tan δ	+2.762	–2.574	+1.003	+0.077	+1.002	–0.066	+1.017	–0.187
Dble. Trans.	May 12		May 20		May 26		May 31	

APPARENT PLACES OF STARS, 2025

TEN-DAY STARS AT UPPER TRANSIT AT GREENWICH

EQUINOX BASED RIGHT ASCENSION – WHOLE NUTATION

FK5-No.	625		635		1456		660	
HIP-No.	82273		83613		84862		86670	
Name	α Trianguli A		60 Herculis		72 Herculis		κ Scorpii	
U.T.	R.A.	Dec.	R.A.	Dec.	R.A.	Dec.	R.A.	Dec.
	$16^{h}51^{m}$	**$-69°04'$**	**$17^{h}06^{m}$**	**$+12°42'$**	**$17^{h}21^{m}$**	**$+32°25'$**	**$17^{h}44^{m}$**	**$-39°02'$**
12 $28^{d}.4$	$15^{s}.9559$	$11''.778$	$30^{s}.9652$	$18''.890$	$34^{s}.2257$	$60''.924$	$11^{s}.2897$	$31''.750$
1 7.4	16.5173	10.066	31.1806	16.415	34.4216	57.661	11.5401	31.134
1 17.4	17.2212	08.592	31.4500	14.185	34.6750	54.728	11.8680	30.500
1 27.4	17.9201	07.231	31.7095	12.357	34.9332	52.268	12.1797	29.800
2 6.3	18.6562	06.510	31.9888	10.564	35.2210	49.952	12.5176	29.477
2 16.3	19.4544	06.361	32.2973	08.953	35.5419	47.952	12.9002	29.425
2 26.3	20.3014	06.360	32.6194	07.954	35.8804	46.705	13.3106	29.224
3 8.2	21.1067	06.719	32.9236	07.432	36.2095	46.077	13.7014	29.111
3 18.2	21.8327	07.644	33.2016	07.124	36.5199	45.798	14.0595	29.341
3 28.2	22.6287	08.970	33.5118	07.188	36.8560	46.049	14.4761	29.661
4 7.2	23.3970	10.401	33.8078	07.912	37.1767	47.110	14.8861	29.835
4 17.1	23.9985	12.058	34.0379	09.010	37.4386	48.647	15.2097	30.128
4 27.1	24.5952	14.258	34.2778	10.133	37.7024	50.333	15.5575	30.772
5 7.1	25.1914	16.602	34.5212	11.616	37.9588	52.493	15.9251	31.388
5 17.1	25.6414	18.812	34.7033	13.533	38.1541	55.146	16.2155	31.858
5 27.0	25.9847	21.331	34.8523	15.388	38.3122	57.791	16.4654	32.618
6 6.0	26.2657	24.104	34.9870	17.109	38.4460	60.344	16.7031	33.635
6 16.0	26.4836	26.650	35.0966	19.024	38.5456	63.101	16.9145	34.482
6 25.9	26.5673	29.041	35.1539	20.992	38.5897	65.898	17.0569	35.292
7 5.9	26.4794	31.476	35.1504	22.640	38.5717	68.326	17.1124	36.355
7 15.9	26.3727	33.811	35.1458	24.033	38.5404	70.451	17.1699	37.461
7 25.9	26.1993	35.671	35.1110	25.491	38.4717	72.578	17.1887	38.278
8 4.8	25.8308	37.096	34.9954	26.790	38.3260	74.427	17.0881	38.991
8 14.8	25.4199	38.427	34.8697	27.571	38.1639	75.651	16.9687	39.854
8 24.8	25.0222	39.321	34.7481	28.147	37.9958	76.566	16.8547	40.502
9 3.7	24.5418	39.439	34.5827	28.728	37.7846	77.340	16.6738	40.695
9 13.7	24.0153	39.223	34.3981	28.904	37.5545	77.568	16.4567	40.839
9 23.7	23.5063	38.743	34.2220	28.587	37.3287	77.164	16.2429	40.977
10 3.7	23.0867	37.618	34.0731	28.145	37.1242	76.501	16.0628	40.684
10 13.6	22.7033	35.946	33.9299	27.568	36.9266	75.571	15.8824	40.056
10 23.6	22.3096	34.022	33.7802	26.549	36.7287	74.055	15.6784	39.423
11 2.6	22.0973	31.962	33.7031	25.089	36.5975	71.997	15.5691	38.744
11 12.6	22.0480	29.497	33.6825	23.603	36.5205	69.835	15.5344	37.706
11 22.5	22.0132	26.744	33.6590	21.990	36.4523	67.447	15.4827	36.512
12 2.5	22.1204	24.268	33.6965	19.880	36.4486	64.511	15.5058	35.583
12 12.5	22.4434	21.900	33.8196	17.616	36.5281	61.419	15.6451	34.648
12 22.4	22.8497	19.399	33.9666	15.506	36.6414	58.469	15.8104	33.506
12 32.4	23.3206	17.226	34.1392	13.297	36.7922	55.442	16.0040	32.550
Mean Place	23.2709	14.263	33.7139	27.315	36.9059	71.076	15.2307	25.624
sec δ, tan δ	+2.799	−2.615	+1.025	+0.225	+1.185	+0.636	+1.287	−0.811
Dble. Trans.	June 4		June 8		June 11		June 17	

APPARENT PLACES OF STARS, 2025

TEN-DAY STARS AT UPPER TRANSIT AT GREENWICH

EQUINOX BASED RIGHT ASCENSION – WHOLE NUTATION

FK5-No.	671		691		700		699	
HIP-No.	87585		90422		90647		91262	
Name	ξ Draconis		α Telescopii		Groombridge 2655		α Lyrae	
U.T.	R.A.	Dec.	R.A.	Dec.	R.A.	Dec.	R.A.	Dec.
	$17^{h}53^{m}$	**+ 56°51′**	**$18^{h}28^{m}$**	**– 45°57′**	**$18^{h}28^{m}$**	**+ 77°33′**	**$18^{h}37^{m}$**	**+ 38°48′**
12 $28^{d}.5$	$55^{s}.2221$	58″.244	$47^{s}.5407$	16″.262	$25^{s}.0015$	44″.442	$45^{s}.2627$	21″.047
1 7.4	55.3708	54.481	47.7538	14.999	24.9930	40.724	45.3631	17.733
1 17.4	55.5984	51.059	48.0548	13.636	25.1621	37.274	45.5310	14.685
1 27.4	55.8769	48.080	48.3441	12.218	25.5608	34.140	45.7205	11.925
2 6.4	56.2152	45.264	48.6726	11.163	26.1382	31.080	45.9493	09.210
2 16.3	56.6038	42.822	49.0598	10.372	26.8393	28.341	46.2231	06.782
2 26.3	57.0302	41.200	49.4835	09.446	27.6687	26.368	46.5347	05.075
3 8.3	57.4745	40.260	49.8963	08.676	28.6213	25.011	46.8567	03.924
3 18.3	57.9186	39.736	50.2864	08.336	29.6391	24.028	47.1738	03.085
3 28.2	58.3780	39.879	50.7480	08.097	30.6456	23.741	47.5298	02.896
4 7.2	58.8237	40.947	51.2094	07.791	31.6521	24.388	47.8902	03.578
4 17.2	59.2246	42.557	51.5867	07.751	32.6552	25.559	48.2073	04.713
4 27.1	59.6049	44.447	51.9996	08.107	33.5526	27.080	48.5307	06.152
5 7.1	59.9502	46.954	52.4386	08.492	34.3168	29.301	48.8557	08.248
5 17.1	60.2305	50.022	52.7975	08.864	34.9924	32.118	49.1315	10.884
5 27.1	60.4546	53.165	53.1191	09.617	35.5283	35.076	49.3708	13.622
6 6.0	60.6203	56.313	53.4315	10.695	35.8690	38.142	49.5820	16.440
6 16.0	60.7239	59.735	53.7143	11.669	36.0384	41.573	49.7605	19.602
6 26.0	60.7550	63.232	53.9222	12.705	36.0584	45.143	49.8832	22.880
7 6.0	60.7063	66.365	54.0372	14.096	35.9092	48.412	49.9350	25.843
7 15.9	60.6058	69.235	54.1511	15.543	35.5566	51.527	49.9623	28.662
7 25.9	60.4432	72.113	54.2165	16.746	35.0433	54.729	49.9475	31.575
8 4.9	60.2015	74.645	54.1472	17.944	34.4199	57.605	49.8484	34.167
8 14.8	59.9168	76.522	54.0528	19.291	33.6418	59.899	49.7153	36.209
8 24.8	59.5960	78.067	53.9546	20.397	32.7121	61.941	49.5630	38.037
9 3.8	59.2293	79.378	53.7722	21.071	31.7174	63.762	49.3604	39.679
9 13.8	58.8374	80.051	53.5424	21.680	30.6633	64.963	49.1237	40.744
9 23.7	58.4342	80.013	53.3073	22.227	29.5339	65.484	48.8730	41.185
10 3.7	58.0396	79.635	53.0951	22.253	28.3722	65.690	48.6317	41.375
10 13.7	57.6560	78.884	52.8714	21.876	27.2396	65.505	48.3877	41.229
10 23.7	57.2850	77.412	52.6147	21.444	26.1611	64.553	48.1294	40.370
11 2.6	56.9706	75.326	52.4518	20.804	25.1127	62.992	47.9218	38.975
11 12.6	56.7135	73.061	52.3596	19.669	24.1500	61.225	47.7622	37.437
11 22.6	56.4986	70.432	52.2432	18.322	23.3549	58.995	47.6079	35.474
12 2.5	56.3614	67.176	52.2067	17.100	22.6945	56.081	47.5069	32.879
12 12.5	56.3117	63.733	52.2937	15.712	22.1582	52.942	47.4845	30.105
12 22.5	56.3315	60.352	52.4076	14.054	21.8359	49.756	47.5016	27.305
12 32.5	56.4258	56.832	52.5551	12.528	21.7329	46.319	47.5596	24.263
Mean Place	58.2469	69.569	51.7649	0 5.719	30.2331	54.231	48.1664	31.569
sec δ, tan δ	+1.830	+1.532	+1.438	–1.034	+4.644	+4.535	+1.283	+0.804
Dble. Trans.	June 20		June 28		June 28		July 1	

APPARENT PLACES OF STARS, 2025

TEN-DAY STARS AT UPPER TRANSIT AT GREENWICH

EQUINOX BASED RIGHT ASCENSION – WHOLE NUTATION

FK5-No.	725		1517		748		1533	
HIP-No.	94834		97290		98495		101101	
Name	ω Aql		56 Sagittarii		ϵ Pavonis		69 Aquilae	
U.T.	R.A.	Dec.	R.A.	Dec.	R.A.	Dec.	R.A.	Dec.
	19ʰ18ᵐ	**+ 11°38′**	**19ʰ47ᵐ**	**– 19°41′**	**20ʰ03ᵐ**	**– 72°50′**	**20ʰ30ᵐ**	**– 2°47′**
12 28.ᵈ5	57.ˢ9419	23.″755	47.ˢ7353	67.″677	22.ˢ8204	42.″618	56.ˢ0853	71.″415
1 7.5	58.0272	21.723	47.8233	67.766	22.8925	39.788	56.1176	72.395
1 17.5	58.1745	19.921	47.9600	67.560	23.1378	36.645	56.2103	73.106
1 27.5	58.3223	18.279	48.1110	67.340	23.4034	33.423	56.2961	73.756
2 6.4	58.4998	16.569	48.2873	67.192	23.8125	30.454	56.4116	74.508
2 16.4	58.7214	15.017	48.5146	67.020	24.3855	27.685	56.5771	75.134
2 26.4	58.9784	14.021	48.7774	66.472	25.0501	24.817	56.7822	75.308
3 8.3	59.2388	13.369	49.0409	65.834	25.7555	22.240	56.9945	75.293
3 18.3	59.4912	12.817	49.2977	65.372	26.5002	20.262	57.2054	75.314
3 28.3	59.7984	12.772	49.6230	64.649	27.4013	18.437	57.4847	74.884
4 7.3	60.1174	13.391	49.9617	63.585	28.3242	16.775	57.7861	73.961
4 17.2	60.3900	14.233	50.2479	62.638	29.1553	15.723	58.0475	73.031
4 27.2	60.6888	15.260	50.5758	61.765	30.0925	15.187	58.3508	71.953
5 7.2	61.0111	16.816	50.9386	60.649	31.0802	14.883	58.6932	70.433
5 17.2	61.2888	18.749	51.2510	59.452	31.9367	14.925	58.9983	68.729
5 27.1	61.5429	20.684	51.5462	58.487	32.7617	15.580	59.2913	67.096
6 6.1	61.7898	22.646	51.8454	57.700	33.5854	16.754	59.5902	65.466
6 16.1	62.0197	24.909	52.1303	56.778	34.3319	18.052	59.8815	63.606
6 26.0	62.2014	27.238	52.3641	55.952	34.9407	19.684	60.1301	61.776
7 6.0	62.3177	29.234	52.5306	55.571	35.4006	21.922	60.3173	60.331
7 16.0	62.4281	31.158	52.6998	55.276	35.8297	24.291	60.5062	58.907
7 26.0	62.5056	33.219	52.8328	54.875	36.1229	26.591	60.6645	57.380
8 4.9	62.4911	34.984	52.8588	54.770	36.1588	29.136	60.7253	56.229
8 14.9	62.4506	36.328	52.8621	54.975	36.1271	31.833	60.7604	55.406
8 24.9	62.4014	37.602	52.8579	55.124	36.0257	34.283	60.7862	54.576
9 3.9	62.2938	38.798	52.7797	55.207	35.7082	36.397	60.7439	53.843
9 13.8	62.1466	39.568	52.6550	55.513	35.2709	38.404	60.6536	53.460
9 23.8	61.9855	39.901	52.5156	56.009	34.7809	40.222	60.5427	53.381
10 3.8	61.8335	40.188	52.3815	56.261	34.2471	41.339	60.4321	53.224
10 13.7	61.6686	40.306	52.2248	56.408	33.6323	41.908	60.2963	53.150
10 23.7	61.4733	39.871	52.0278	56.813	32.9358	42.257	60.1158	53.526
11 2.7	61.3300	39.138	51.8896	57.153	32.3560	42.006	59.9798	53.967
11 12.7	61.2296	38.436	51.7929	57.155	31.8384	40.952	59.8769	54.233
11 22.6	61.1101	37.412	51.6628	57.211	31.2628	39.497	59.7382	54.769
12 2.6	61.0353	35.908	51.5828	57.442	30.8372	37.771	59.6357	55.586
12 12.6	61.0381	34.359	51.5892	57.463	30.6115	35.479	59.6060	56.269
12 22.6	61.0610	32.815	51.6074	57.299	30.4305	32.701	59.5848	56.915
12 32.5	61.1062	31.031	51.6460	57.239	30.3514	29.824	59.5778	57.747
Mean Place	60.8183	35.332	50.8332	52.055	29.7055	21.690	58.8797	56.786
sec δ, tan δ	+1.021	+0.206	+1.062	–0.358	+3.389	–3.238	+1.001	–0.049
Dble. Trans.	July 11		July 18		July 22		July 29	

APPARENT PLACES OF STARS, 2025

TEN-DAY STARS AT UPPER TRANSIT AT GREENWICH

EQUINOX BASED RIGHT ASCENSION – WHOLE NUTATION

FK5-No.	777		1546		1577		855	
HIP-No.	102098		102978		108036		112029	
Name	α Cygni		ω Capricorni		μ Capricorni		ζ Pegasi	
U.T.	R.A.	Dec.	R.A.	Dec.	R.A.	Dec.	R.A.	Dec.
	20h42m	**+ 45°21′**	**20h53m**	**– 26°49′**	**21h54m**	**– 13°25′**	**22h42m**	**+ 10°57′**
12 28.d6	15.s0474	75.″219	17.s3893	39.″226	38.s5492	68.″417	41.s8098	44.″337
1 7.6	14.9724	72.412	17.4116	38.821	38.5179	68.701	41.7325	43.253
1 17.5	14.9714	69.665	17.4987	38.058	38.5408	68.636	41.7036	42.325
1 27.5	15.0034	66.808	17.5719	37.217	38.5501	68.542	41.6632	41.209
2 6.5	15.0765	63.779	17.6808	36.374	38.5897	68.432	41.6413	40.014
2 16.5	15.2063	60.891	17.8507	35.438	38.6626	68.301	41.6635	38.979
2 26.4	15.4045	58.494	18.0603	34.152	38.8017	67.484	41.7318	38.246
3 8.4	15.6396	56.409	18.2773	32.822	38.9429	66.683	41.8151	37.602
3 18.4	15.8880	54.496	18.4985	31.696	39.0907	65.961	41.9054	36.961
3 28.3	16.2045	53.277	18.7987	30.283	39.3170	64.791	42.0760	36.881
4 7.3	16.5661	52.850	19.1211	28.615	39.5745	63.273	42.2905	37.247
4 17.3	16.9111	52.755	19.4011	27.194	39.8025	61.908	42.4856	37.610
4 27.3	17.2804	53.166	19.7371	25.825	40.0883	60.375	42.7357	38.390
5 7.2	17.6828	54.410	20.1190	24.261	40.4276	58.483	43.0471	39.753
5 17.2	18.0659	56.194	20.4582	22.772	40.7404	56.608	43.3479	41.276
5 27.2	18.4240	58.271	20.7919	21.566	41.0558	54.850	43.6535	42.950
6 6.2	18.7626	60.726	21.1410	20.566	41.3929	53.118	43.9809	44.900
6 16.1	19.0874	63.722	21.4816	19.524	41.7330	51.275	44.3225	47.165
6 26.1	19.3673	66.981	21.7765	18.691	42.0402	49.587	44.6412	49.478
7 6.1	19.5695	70.125	22.0104	18.388	42.2963	48.349	44.9106	51.591
7 16.0	19.7434	73.455	22.2510	18.181	42.5632	47.126	45.1912	53.899
7 26.0	19.8813	77.055	22.4546	17.968	42.8036	45.919	45.4566	56.320
8 5.0	19.9257	80.378	22.5487	18.194	42.9477	45.227	45.6352	58.324
8 15.0	19.9139	83.418	22.6197	18.722	43.0692	44.816	45.7864	60.185
8 24.9	19.8721	86.480	22.6787	19.210	43.1812	44.409	45.9306	62.109
9 3.9	19.7712	89.352	22.6536	19.747	43.2178	44.226	46.0114	63.757
9 13.9	19.6108	91.736	22.5744	20.519	43.2007	44.378	46.0370	65.050
9 23.9	19.4064	93.654	22.4730	21.446	43.1572	44.760	46.0297	66.093
10 3.8	19.1945	95.391	22.3641	22.108	43.1027	45.044	46.0136	67.095
10 13.8	18.9607	96.748	22.2193	22.664	43.0098	45.405	45.9607	67.840
10 23.8	18.6784	97.328	22.0229	23.449	42.8595	46.148	45.8439	68.044
11 2.7	18.4158	97.470	21.8748	24.013	42.7420	46.765	45.7464	68.201
11 12.7	18.1868	97.408	21.7548	24.156	42.6424	47.121	45.6641	68.364
11 22.7	17.9401	96.661	21.5883	24.331	42.4903	47.713	45.5263	68.010
12 2.7	17.7138	95.231	21.4647	24.500	42.3627	48.352	45.3940	67.403
12 12.6	17.5500	93.578	21.4202	24.291	42.2969	48.668	45.3102	66.897
12 22.6	17.4193	91.587	21.3777	23.847	42.2249	48.903	45.2179	66.169
12 32.6	17.3160	89.089	21.3509	23.411	42.1569	49.202	45.1186	65.149
Mean Place	18.1187	80.930	20.2874	19.404	41.0042	50.289	44.0854	54.444
sec δ, tan δ	+1.423	+1.013	+1.121	–0.506	+1.028	–0.239	+1.019	+0.194
Dble. Trans.	August 1		August 4		August 20		September 1	

APPARENT PLACES OF STARS, 2025

NORTHERN CIRCUMPOLAR STARS AT UPPER TRANSIT AT GREENWICH

EQUINOX BASED RIGHT ASCENSION – WHOLE NUTATION

FK6 Star No. 907 = Hipparcos Star No. 11767 = α Ursae Minoris (Polaris)

Day	January		February		March		April		May		June	
	R.A.	Dec.	R.A.	Dec.	R.A.	Dec.	R.A.	Dec.	R.A.	Dec.	R.A.	Dec.
	03^h04^m	$+89°22'$	03^h03^m	$+89°22'$	03^h02^m	$+89°22'$	03^h02^m	$+89°22'$	03^h02^m	$+89°22'$	03^h02^m	$+89°22'$
1	63.497	26.543	69.702	31.449	76.436	30.528	32.989	24.164	20.741	15.417	40.924	06.841
2	61.893	26.830	67.581	31.465	74.573	30.354	32.440	23.891	21.031	15.165	41.814	06.615
3	60.166	27.095	65.623	31.461	72.922	30.167	31.875	23.642	21.188	14.916	42.726	06.372
4	58.386	27.329	63.837	31.449	71.456	29.984	31.223	23.412	21.237	14.659	43.709	06.112
5	56.635	27.531	62.187	31.445	70.107	29.818	30.449	23.189	21.225	14.386	44.799	05.841
6	54.987	27.706	60.608	31.457	68.786	29.673	29.558	22.962	21.208	14.094	46.017	05.566
7	53.483	27.867	59.018	31.488	67.416	29.546	28.579	22.721	21.238	13.782	47.368	05.293
8	52.126	28.025	57.346	31.535	65.940	29.431	27.558	22.462	21.359	13.455	48.841	05.030
9	50.876	28.195	55.541	31.591	64.338	29.318	26.548	22.181	21.602	13.116	50.413	04.783
10	49.661	28.384	53.587	31.645	62.616	29.195	25.598	21.878	21.985	12.773	52.048	04.556
11	48.395	28.595	51.496	31.686	60.807	29.056	24.751	21.558	22.508	12.432	53.701	04.352
12	47.003	28.822	49.309	31.705	58.958	28.894	24.034	21.225	23.155	12.100	55.325	04.169
13	45.441	29.055	47.080	31.698	57.124	28.706	23.459	20.886	23.899	11.784	56.882	04.005
14	43.702	29.282	44.863	31.665	55.354	28.495	23.023	20.550	24.699	11.488	58.344	03.852
15	41.814	29.493	42.708	31.607	53.688	28.264	22.705	20.222	25.511	11.211	59.705	03.701
16	39.828	29.680	40.647	31.530	52.149	28.020	22.473	19.908	26.290	10.953	60.988	03.542
17	37.802	29.840	38.700	31.440	50.744	27.770	22.286	19.611	27.003	10.710	62.236	03.367
18	35.788	29.974	36.865	31.346	49.464	27.522	22.100	19.331	27.628	10.473	63.517	03.173
19	33.829	30.086	35.131	31.254	48.284	27.281	21.874	19.066	28.165	10.234	64.903	02.959
20	31.950	30.182	33.471	31.169	47.170	27.052	21.574	18.811	28.642	09.983	66.459	02.732
21	30.165	30.269	31.847	31.096	46.080	26.838	21.184	18.558	29.111	09.712	68.220	02.506
22	28.470	30.354	30.215	31.038	44.971	26.639	20.709	18.298	29.645	09.419	70.176	02.293
23	26.847	30.443	28.525	30.993	43.799	26.452	20.185	18.019	30.326	09.104	72.266	02.109
24	25.266	30.543	26.731	30.956	42.531	26.269	19.682	17.713	31.219	08.777	74.394	01.960
25	23.683	30.656	24.807	30.918	41.155	26.083	19.289	17.380	32.345	08.455	76.459	01.844
26	22.046	30.785	22.754	30.867	39.690	25.879	19.093	17.024	33.666	08.153	78.389	01.751
27	20.302	30.925	20.616	30.789	38.194	25.649	19.143	16.661	35.088	07.884	80.160	01.666
28	18.412	31.069	18.477	30.677	36.758	25.385	19.425	16.308	36.500	07.649	81.796	01.576
29	16.362	31.204	16.436	30.528	35.481	25.090	19.857	15.981	37.810	07.440	83.350	01.471
30	14.179	31.319			34.433	24.776	20.329	15.686	38.973	07.245	84.888	01.348
31	11.930	31.402			33.621	24.463	20.741	15.417	39.997	07.049	86.469	01.208
32	09.702	31.449			32.989	24.164			40.924	06.841		
	sec(δ)	tan(δ)	sec(δ)	tan(δ)	sec(δ)	tan(δ)	sec(δ)	tan(δ)	sec(δ)	tan(δ)	sec(δ)	tan(δ)
	91.66	91.66	91.74	91.73	91.59	91.59	91.27	91.26	90.91	90.90	90.61	90.60

Mean R.A. $03^h05^m10^s.998$ Double lower transit May 7 Mean Dec. $+89°22'10''.295$

Apparent places of Northern Circumpolar Stars

APPARENT PLACES OF STARS, 2025

NORTHERN CIRCUMPOLAR STARS AT UPPER TRANSIT AT GREENWICH

EQUINOX BASED RIGHT ASCENSION – WHOLE NUTATION

FK6 Star No. 907 = Hipparcos Star No. 11767 = α Ursae Minoris (Polaris)

Day	July		August		September		October		November		December	
	R.A.	Dec.	R.A.	Dec.	R.A.	Dec.	R.A.	Dec.	R.A.	Dec.	R.A.	Dec.
	03^h03^m	$+89°21'$	03^h04^m	$+89°21'$	03^h05^m	$+89°22'$	03^h06^m	$+89°22'$	03^h07^m	$+89°22'$	03^h06^m	$+89°22'$
1	26.469	61.208	29.768	59.630	36.905	02.925	32.311	10.269	07.926	20.770	73.659	31.936
2	28.135	61.056	32.088	59.627	39.141	03.131	33.790	10.615	08.318	21.125	73.512	32.232
3	29.913	60.897	34.474	59.643	41.285	03.357	35.126	10.960	08.758	21.456	73.495	32.545
4	31.814	60.740	36.888	59.681	43.304	03.597	36.342	11.293	09.326	21.768	73.515	32.887
5	33.831	60.591	39.288	59.743	45.182	03.843	37.496	11.607	10.067	22.075	73.453	33.260
6	35.944	60.457	41.627	59.829	46.930	04.084	38.668	11.895	10.968	22.395	73.209	33.656
7	38.122	60.343	43.867	59.933	48.593	04.308	39.947	12.162	11.948 12.892	22.743 23.124	72.736	34.059
8	40.325	60.252	45.980	60.048	50.244	04.510	41.392	12.420	13.694	23.534	72.046	34.452
9	42.505	60.186	47.964	60.163	51.970	04.689	43.011	12.684	14.290	23.959	71.196	34.825
10	44.618	60.141	49.848	60.268	53.841	04.854	44.748	12.971	14.672	24.386	70.256	35.170
11	46.630	60.111	51.692	60.355	55.886	05.019	46.506	13.290	14.873	24.801	69.293	35.489
12	48.525	60.088	53.573	60.420	58.081	05.199	48.183	13.641	14.950	25.196	68.359	35.786
13	50.316	60.060	55.565	60.467	60.358	05.406	49.703	14.015	14.965	25.569	67.484	36.068
14	52.042	60.019	57.720	60.506	62.626	05.645	51.031	14.401	14.975	25.920	66.682	36.341
15	53.767	59.959	60.046	60.552	64.802	05.914	52.175	14.786	15.023	26.255	65.949	36.614
16	55.566	59.879	62.508	60.617	66.827	06.204	53.171	15.159	15.134	26.579	65.264	36.892
17	57.503	59.785	65.038	60.712	68.678	06.504	54.073	15.515	15.317	26.900	64.598	37.181
18	59.618	59.688	67.549	60.839	70.367	06.801	54.938	15.852	15.563	27.226	63.913	37.483
19	61.912	59.601	69.964	60.996	71.932	07.087	55.819	16.171	15.849	27.561	63.165	37.800
20	64.343	59.538	72.230	61.172	73.431	07.357	56.753	16.477	16.139	27.911	62.316	38.128
21	66.835	59.507	74.331	61.355	74.920	07.607	57.763	16.777	16.394	28.278	61.336	38.463
22	69.295	59.509	76.286	61.534	76.450	07.843	58.852	17.079	16.572	28.661	60.209	38.798
23	71.644	59.540	78.144	61.698	78.056	08.069	60.004	17.388	16.636	29.057	58.939	39.124
24	73.837	59.586	79.967	61.845	79.751	08.294	61.191	17.713	16.562	29.460	57.552	39.433
25	75.871	59.634	81.814	61.975	81.532	08.524	62.377	18.055	16.337	29.864	56.093	39.718
26	77.785	59.673	83.731	62.093	83.377	08.766	63.519	18.416	15.972	30.260	54.624	39.977
27	79.639	59.695	85.742	62.204	85.254	09.027	64.579	18.796	15.494	30.640	53.211	40.210
28	81.496	59.699	87.853	62.318	87.127	09.307	65.522	19.190	14.955	30.998	51.914	40.424
29	83.412	59.688	90.055	62.442	88.952	09.610	66.325	19.593	14.423	31.331	50.766	40.630
30	85.421	59.668	92.321	62.582	90.691	09.932	66.981	19.996	13.970	31.640	49.763	40.842
31	87.540	59.646	94.617	62.742	92.311	10.269	67.502	20.392	13.659	31.936	48.850	41.073
32	89.768	59.630	96.905	62.925			67.926	20.770			47.931	41.332
	$\sec(\delta)$	$\tan(\delta)$	$\sec(\delta)$	$\tan(\delta)$	$\sec(\delta)$	$\tan(\delta)$	$\sec(\delta)$	$\tan(\delta)$	$\sec(\delta)$	$\tan(\delta)$	$\sec(\delta)$	$\tan(\delta)$
	90.46	90.46	90.49	90.49	90.72	90.71	91.07	91.07	91.54	91.53	91.96	91.95

Mean R.A. $03^h05^m10^s.998$ Double lower transit May 7 Mean Dec. $+89°22'10''.295$

APPARENT PLACES OF STARS, 2025

NORTHERN CIRCUMPOLAR STARS AT UPPER TRANSIT AT GREENWICH

EQUINOX BASED RIGHT ASCENSION – WHOLE NUTATION

FK6 Star No. 1644 = Hipparcos Star No. 72573 = Grb 2196 UMi

Day	January		February		March		April		May		June	
	R.A.	Dec.	R.A.	Dec.	R.A.	Dec.	R.A.	Dec.	R.A.	Dec.	R.A.	Dec.
	14^h48^m	$+82°23'$	14^h48^m	$+82°23'$	14^h48^m	$+82°23'$	14^h49^m	$+82°24'$	14^h48^m	$+82°24'$	14^h48^m	$+82°24'$
1	47.030	63.927	51.986	58.950	56.610	59.989	00.278	06.593	61.355	15.626	59.750	24.636
2	47.183	63.631	52.170	58.927	56.766	60.163	00.331	06.886	61.338	15.891	59.676	24.876
3	47.343	63.355	52.341	58.930	56.907	60.356	00.384	07.154	61.330	16.147	59.595	25.134
4	47.506	63.106	52.501	58.945	57.035	60.550	00.444	07.399	61.329 61.330	16.408 16.682	59.506	25.405
5	47.667	62.890	52.651	58.958	57.155	60.732	00.513	07.634	61.331	16.974	59.407	25.683
6	47.820	62.703	52.797	58.958	57.273	60.894	00.589	07.868	61.326	17.286	59.299	25.960
7	47.962	62.536	52.944	58.940	57.395	61.036	00.670	08.112	61.314	17.616	59.182	26.230
8	48.095	62.377	53.098	58.904	57.525	61.164	00.752	08.373	61.292	17.959	59.060	26.487
9	48.221	62.210	53.261	58.856	57.663	61.286	00.832	08.656	61.260	18.308	58.933	26.724
10	48.346	62.028	53.434	58.806	57.809	61.413	00.906	08.960	61.218	18.658	58.807	26.939
11	48.475	61.824	53.615	58.766	57.959	61.555	00.972	09.284	61.168	19.001	58.683	27.132
12	48.614	61.602	53.802	58.744	58.111	61.718	01.028	09.623	61.111	19.330	58.564	27.304
13	48.765	61.370	53.991	58.748	58.260	61.905	01.073	09.969	61.052	19.641	58.452	27.463
14	48.928	61.139	54.177	58.779	58.403	62.118	01.109	10.316	60.992	19.932	58.346	27.616
15	49.101	60.922	54.357	58.835	58.537	62.351	01.137	10.657	60.935	20.203	58.245	27.773
16	49.279	60.726	54.530	58.913	58.662	62.601	01.159	10.986	60.883	20.458	58.145	27.942
17	49.458	60.558	54.695	59.006	58.778	62.859	01.180	11.299	60.837	20.704	58.042	28.129
18	49.636	60.416	54.851	59.106	58.885	63.118	01.201	11.595	60.796	20.948	57.931	28.336
19	49.808	60.298	55.001	59.207	58.985	63.372	01.225	11.875	60.758	21.199	57.807	28.559
20	49.975	60.199	55.147	59.303	59.082	63.615	01.254	12.142	60.720	21.467	57.669	28.787
21	50.134	60.111	55.291	59.388	59.177	63.845	01.288	12.405	60.676	21.757	57.518	29.006
22	50.289	60.028	55.437	59.460	59.275	64.059	01.328	12.671	60.621	22.070	57.359	29.202
23	50.439	59.943	55.587	59.518	59.377	64.261	01.370	12.951	60.551	22.398	57.197	29.364
24	50.587	59.851	55.745	59.566	59.486	64.454	01.409	13.254	60.465	22.728	57.041	29.491
25	50.737	59.745	55.911	59.611	59.601	64.649	01.439	13.585	60.366	23.044	56.894	29.590
26	50.891	59.625	56.085	59.665	59.721	64.854	01.456	13.941	60.261	23.332	56.758	29.676
27	51.053	59.493	56.263	59.740	59.842	65.083	01.455	14.311	60.156	23.584	56.631	29.761
28	51.226	59.353	56.440	59.846	59.958	65.343	01.438	14.677	60.059	23.806	56.508	29.858
29	51.409	59.218	56.610	59.989	60.062	65.636	01.411	15.024	59.971	24.009	56.385	29.972
30	51.600	59.098			60.149	65.953	01.381	15.340	59.893	24.207	56.258	30.104
31	51.794	59.007			60.220	66.278	01.355	15.626	59.821	24.414	56.123	30.250
32	51.986	58.950			60.278	66.593			59.750	24.636		
	$\sec(\delta)$	$\tan(\delta)$	$\sec(\delta)$	$\tan(\delta)$	$\sec(\delta)$	$\tan(\delta)$	$\sec(\delta)$	$\tan(\delta)$	$\sec(\delta)$	$\tan(\delta)$	$\sec(\delta)$	$\tan(\delta)$
	7.56	7.49	7.56	7.49	7.56	7.50	7.56	7.50	7.57	7.50	7.57	7.50

Mean R.A. $14^h48^m47^s.635$ Double lower transit November 2 Mean Dec. $+82°24'19''.880$

APPARENT PLACES OF STARS, 2025

NORTHERN CIRCUMPOLAR STARS AT UPPER TRANSIT AT GREENWICH

EQUINOX BASED RIGHT ASCENSION – WHOLE NUTATION

FK6 Star No. 1644 = Hipparcos Star No. 72573 = Grb 2196 UMi

Day	July		August		September		October		November		December	
	R.A.	Dec.	R.A.	Dec.	R.A.	Dec.	R.A.	Dec.	R.A.	Dec.	R.A.	Dec.
	14^h48^m	$+82°24'$	14^h48^m	$+82°24'$	14^h48^m	$+82°24'$	14^h48^m	$+82°24'$	14^h48^m	$+82°23'$	14^h48^m	$+82°23'$
1	56.123	30.250	51.116	31.553	45.944	27.722	41.898	19.738	39.683	68.622	39.992	57.323
2	55.980	30.405	50.930	31.535	45.784	27.487	41.805	19.373	39.665	68.264	40.032	57.017
3	55.827	30.561	50.744	31.495	45.633	27.235	41.720	19.016	39.640	67.928	40.065	56.699
4	55.667	30.711	50.560	31.432	45.493	26.974	41.639	18.675	39.604	67.602	40.097	56.357
5	55.501	30.848	50.381	31.344	45.360	26.714	41.557	18.357	39.558	67.269	40.136	55.983
6	55.332	30.966	50.210	31.236	45.233	26.467	41.467	18.063	39.508	66.914	40.189	55.582
7	55.162	31.062	50.048	31.114	45.106	26.239	41.365	17.783	39.461	66.526	40.258	55.169
8	54.994	31.134	49.894	30.987	44.974	26.034	41.253	17.502	39.427	66.108	40.342	54.760
9	54.832	31.182	49.747	30.866	44.831	25.847	41.134	17.204	39.407	65.668	40.436	54.370
10	54.677	31.213	49.602	30.761	44.677	25.666	41.014	16.876	39.403	65.223	40.536	54.005
11	54.530	31.234	49.453	30.676	44.512	25.475	40.902	16.516	39.411	64.786	40.636	53.669
12	54.390	31.255	49.296	30.611	44.343	25.260	40.802	16.128	39.428	64.366	40.732	53.356
13	54.253	31.285	49.127	30.557	44.177	25.014	40.716	15.725	39.447	63.969	40.825	53.062
14	54.115	31.333	48.947	30.502	44.018	24.737	40.643	15.318	39.466	63.595	40.912	52.779
15	53.971	31.400	48.757	30.432	43.870	24.435	40.580	14.920	39.482	63.239	40.995	52.499
16	53.817	31.483	48.565	30.336	43.735	24.119	40.523	14.538	39.493	62.896	41.075	52.216
17	53.650	31.574	48.375	30.207	43.612	23.801	40.468	14.175	39.500	62.559	41.156	51.924
18	53.472	31.660	48.192	30.047	43.496	23.492	40.411	13.830	39.503	62.220	41.240	51.619
19	53.284	31.726	48.021	29.863	43.384	23.198	40.349	13.501	39.505	61.874	41.330	51.301
20	53.094	31.763	47.862	29.668	43.272	22.922	40.283	13.180	39.509	61.514	41.427	50.971
21	52.906	31.766	47.712	29.472	43.156	22.663	40.211	12.860	39.516	61.138	41.535	50.632
22	52.727	31.738	47.568	29.288	43.035	22.415	40.137	12.534	39.531	60.746	41.653	50.292
23	52.559	31.689	47.425	29.120	42.907	22.172	40.061	12.196	39.555	60.340	41.781	49.958
24	52.402	31.632	47.280	28.970	42.774	21.925	39.987	11.840	39.590	59.925	41.916	49.639
25	52.253	31.581	47.128	28.834	42.638	21.668	39.919	11.466	39.635	59.508	42.054	49.341
26	52.106	31.544	46.970	28.706	42.501	21.395	39.857	11.074	39.690	59.096	42.193	49.069
27	51.958	31.525	46.805	28.579	42.366	21.102	39.805	10.665	39.752	58.697	42.328	48.823
28	51.805	31.522	46.634	28.444	42.236	20.787	39.764	10.247	39.817	58.319	42.454	48.599
29	51.644	31.531	46.459	28.294	42.114	20.453	39.733	09.825	39.882	57.965	42.571	48.388
30	51.475	31.544	46.284	28.126	42.001	20.101	39.712	09.408	39.942	57.635	42.680	48.177
31	51.298	31.554	46.112	27.936	41.898	19.738	39.696	09.004	39.992	57.323	42.783	47.951
32	51.116	31.553	45.944	27.722			39.683	08.622			42.889	47.700
	sec(δ)	tan(δ)	sec(δ)	tan(δ)	sec(δ)	tan(δ)	sec(δ)	tan(δ)	sec(δ)	tan(δ)	sec(δ)	tan(δ)
	7.57	7.50	7.57	7.50	7.57	7.50	7.57	7.50	7.56	7.50	7.56	7.49

Mean R.A. $14^h48^m47^s.635$ Double lower transit November 2 Mean Dec. $+82°24'19''.880$

APPARENT PLACES OF STARS, 2025

SOUTHERN CIRCUMPOLAR STARS AT UPPER TRANSIT AT GREENWICH

EQUINOX BASED RIGHT ASCENSION – WHOLE NUTATION

FK6 Star No. 1662 = Hipparcos Star No. 30678 = A Octantis

Day	January		February		March		April		May		June	
	R.A.	Dec.	R.A.	Dec.	R.A.	Dec.	R.A.	Dec.	R.A.	Dec.	R.A.	Dec.
	06^h03^m	$-88°45'$	06^h02^m	$-88°45'$	06^h02^m	$-88°45'$	06^h01^m	$-88°45'$	06^h01^m	$-88°45'$	06^h01^m	$-88°44'$
1	39.664	04.234	78.041	12.819	48.118	17.447	70.983	18.262	37.731	14.547	13.464	66.848
2	39.191	04.501	77.236	13.055	47.061	17.572	69.667	18.248	36.617	14.328	12.993	66.514
3	38.767	04.770	76.387	13.314	45.945	17.715	68.322	18.203	35.571	14.086	12.583	66.195
4	38.381	05.053	75.466	13.587	44.750	17.863	66.987	18.125	34.604	13.832	12.210	65.893
5	38.007	05.359	74.461	13.859	43.477	17.998	65.694	18.018	33.715	13.577	11.853	65.609
6	37.610	05.691	73.379	14.116	42.151	18.109	64.463	17.892	32.891	13.329	11.492	65.341
7	37.156	06.046	72.242	14.349	40.802	18.189	63.301	17.756	32.113	13.095	11.112	65.083
8	36.619	06.413	71.082	14.550	39.463	18.237	62.205	17.621	31.362	12.876	10.701	64.830
9	35.990	06.778	69.931	14.722	38.162	18.260	61.161	17.493	30.616	12.673	10.257	64.575
10	35.279	07.127	68.816	14.871	36.917	18.266	60.151	17.379	29.856	12.483	09.783	64.310
11	34.513	07.449	67.752	15.005	35.731	18.264	59.155	17.279	29.068	12.299	09.290	64.028
12	33.727	07.740	66.742	15.137	34.600	18.265	58.151	17.194	28.245	12.117	08.796	63.725
13	32.956	08.004	65.777	15.274	33.511	18.276	57.123	17.119	27.385	11.927	08.322	63.400
14	32.226	08.249	64.840	15.424	32.442	18.301	56.059	17.048	26.495	11.724	07.892	63.055
15	31.546	08.487	63.908	15.590	31.372	18.341	54.952	16.974	25.587	11.500	07.524	62.698
16	30.912	08.727	62.960	15.770	30.283	18.393	53.803	16.890	24.682	11.254	07.230	62.336
17	30.310	08.978	61.977	15.962	29.156	18.453	52.622	16.788	23.799	10.984	07.010	61.983
18	29.720	09.244	60.946	16.160	27.983	18.515	51.422	16.664	22.962	10.694	06.848	61.647
19	29.119	09.527	59.858	16.357	26.760	18.571	50.223	16.516	22.187	10.391	06.719	61.337
20	28.488	09.826	58.710	16.546	25.489	18.614	49.048	16.342	21.484	10.085	06.586	61.055
21	27.810	10.135	57.508	16.720	24.181	18.637	47.918	16.148	20.850	09.788	06.416	60.794
22	27.070	10.450	56.263	16.871	22.851	18.636	46.851	15.942	20.267	09.511	06.183	60.542
23	26.264	10.763	54.995	16.996	21.521	18.608	45.854	15.736	19.702	09.260	05.886	60.283
24	25.390	11.065	53.731	17.092	20.216	18.553	44.921	15.543	19.115	09.036	05.546	60.001
25	24.456	11.350	52.498	17.164	18.960	18.480	44.023	15.374	18.470	08.829	05.203	59.688
26	23.479	11.611	51.321	17.221	17.770	18.398	43.120	15.234	17.751	08.622	04.902	59.344
27	22.485	11.843	50.210	17.278	16.648	18.322	42.168	15.116	16.970	08.396	04.677	58.979
28	21.503	12.050	49.153	17.350	15.574	18.267	41.138	15.006	16.163	08.138	04.545	58.606
29	20.561	12.238	48.118	17.447	14.510	18.241	40.034	14.883	15.377	07.846	04.502	58.237
30	19.675	12.419			13.409	18.240	38.884	14.732	14.653	07.526	04.533	57.883
31	18.843	12.609			12.236	18.253	37.731	14.547	14.014	07.189	04.614	57.549
32	18.041	12.819			10.983	18.262			13.464	06.848		
	sec(δ)	tan(δ)	sec(δ)	tan(δ)	sec(δ)	tan(δ)	sec(δ)	tan(δ)	sec(δ)	tan(δ)	sec(δ)	tan(δ)
	45.93	45.92	46.00	45.99	46.03	46.02	46.01	46.00	45.96	45.94	45.86	45.85

Mean R.A. $06^h02^m08^s.735$ Double lower transit June 21 Mean Dec. $-88°45'09''.191$

APPARENT PLACES OF STARS, 2025

SOUTHERN CIRCUMPOLAR STARS AT UPPER TRANSIT AT GREENWICH

EQUINOX BASED RIGHT ASCENSION – WHOLE NUTATION

FK6 Star No. 1662 = Hipparcos Star No. 30678 = A Octantis

Day	July		August		September		October		November		December	
	R.A.	Dec.	R.A.	Dec.	R.A.	Dec.	R.A.	Dec.	R.A.	Dec.	R.A.	Dec.
	06^h01^m	$-88°44'$	06^h01^m	$-88°44'$	06^h01^m	$-88°44'$	06^h01^m	$-88°44'$	06^h02^m	$-88°44'$	06^h02^m	$-88°44'$
1	04.614	57.549	11.855	48.462	32.721	42.386	59.847	41.169	25.805	45.420	39.702	53.956
2	04.721	57.237	12.362	48.249	33.458	42.268	60.708	41.181	26.601	45.623	39.969	54.329
3	04.832	56.944	12.833	48.036	34.199	42.134	61.616	41.189	27.402	45.859	40.141	54.721
4	04.929	56.666	13.276	47.815	34.967	41.986	62.579	41.203	28.169	46.130	40.203	55.117
5	05.000	56.398	13.699	47.582	35.784	41.829	63.595	41.238	28.860	46.430	40.166	55.496
6	05.039	56.131	14.122	47.332	36.666	41.672	64.643	41.304	29.448	46.744	40.067	55.846
7	05.047	55.859	14.568	47.064	37.616	41.529	65.685	41.407	29.935	47.052	39.958	56.162
8	05.031	55.576	15.059	46.782	38.621	41.411	66.678	41.546	30.352	47.338	39.885	56.453
9	05.007	55.275	15.616	46.493	39.649	41.329	67.590	41.706	30.744	47.595	39.870	56.732
10	04.995	54.953	16.248	46.208	40.661	41.282	68.412	41.871	31.157	47.827	39.916	57.014
11	05.020	54.613	16.952	45.941	41.621	41.263	69.161	42.023	31.619	48.044	40.006	57.310
12	05.102	54.259	17.703	45.702	42.508	41.258	69.874	42.153	32.140	48.259	40.116	57.627
13	05.259	53.901	18.470	45.496	43.325	41.250	70.590	42.258	32.714	48.485	40.220	57.966
14	05.492	53.549	19.215	45.320	44.096	41.226	71.345	42.346	33.323	48.729	40.294	58.325
15	05.791	53.216	19.909	45.165	44.853	41.180	72.156	42.426	33.946	48.997	40.320	58.700
16	06.130	52.909	20.541	45.016	45.632	41.113	73.029	42.510	34.557	49.289	40.285	59.085
17	06.477	52.632	21.117	44.860	46.462	41.031	73.956	42.608	35.136	49.603	40.182	59.471
18	06.797	52.380	21.661	44.684	47.359	40.946	74.918	42.729	35.664	49.934	40.013	59.853
19	07.065	52.143	22.208	44.485	48.326	40.867	75.894	42.875	36.131	50.276	39.785	60.223
20	07.273	51.906	22.793	44.264	49.351	40.805	76.859	43.048	36.533	50.620	39.512	60.576
21	07.430	51.656	23.444	44.030	50.417	40.769	77.793	43.244	36.873	50.961	39.212 38.904	60.910 61.225
22	07.567	51.381	24.175	43.793	51.498	40.759	78.681	43.457	37.160	51.291	38.609	61.525
23	07.724	51.078	24.984	43.566	52.571	40.776	79.511	43.681	37.409	51.607	38.339	61.815
24	07.938	50.752	25.857	43.359	53.616	40.816	80.284	43.908	37.638	51.906	38.102	62.106
25	08.235	50.414	26.770	43.178	54.619	40.871	81.005	44.131	37.867	52.190	37.895	62.405
26	08.622	50.078	27.698	43.023	55.574	40.935	81.683	44.345	38.113	52.464	37.702	62.722
27	09.090	49.754	28.618	42.893	56.481	41.001	82.336	44.545	38.387	52.733	37.499	63.063
28	09.617	49.451	29.512	42.781	57.348	41.061	82.980	44.732	38.695	53.008	37.253	63.429
29	10.180	49.172	30.370	42.682	58.187	41.110	83.635	44.906	39.031	53.298	36.933	63.814
30	10.753	48.918	31.189	42.588	59.013	41.146	84.318	45.073	39.377	53.612	36.513	64.206
31	11.317	48.683	31.969	42.491	59.847	41.169	85.040	45.241	39.702	53.956	35.990	64.589
32	11.855	48.462	32.721	42.386			85.805	45.420			35.385	64.948
	$\sec(\delta)$	$\tan(\delta)$	$\sec(\delta)$	$\tan(\delta)$	$\sec(\delta)$	$\tan(\delta)$	$\sec(\delta)$	$\tan(\delta)$	$\sec(\delta)$	$\tan(\delta)$	$\sec(\delta)$	$\tan(\delta)$
	45.77	45.76	45.69	45.68	45.65	45.64	45.66	45.65	45.73	45.72	45.83	45.82

Mean R.A. $06^h02^m08^s.735$ Double lower transit June 21 Mean Dec. $-88°45'09''.191$

APPARENT PLACES OF STARS, 2025

SOUTHERN CIRCUMPOLAR STARS AT UPPER TRANSIT AT GREENWICH

EQUINOX BASED RIGHT ASCENSION – WHOLE NUTATION

FK6 Star No. 1666 = Hipparcos Star No. 76996 = ρ Oct

Day	January		February		March		April		May		June	
	R.A.	Dec.	R.A.	Dec.	R.A.	Dec.	R.A.	Dec.	R.A.	Dec.	R.A.	Dec.
	15^h49^m	$-84°32'$	15^h49^m	$-84°32'$	15^h49^m	$-84°32'$	15^h49^m	$-84°32'$	15^h49^m	$-84°32'$	15^h49^m	$-84°32'$
1	12.149	21.713	19.953	18.926	27.704	20.638	35.338	26.556	40.654	35.292	42.894	46.026
2	12.366	21.596	20.193	18.898	27.944	20.737	35.597	26.781	40.819	35.653	42.863	46.366
3	12.569	21.475	20.447	18.852	28.203	20.823	35.860	27.038	40.960	36.029	42.826	46.682
4	12.761	21.340	20.722	18.800	28.484	20.910	36.114	27.324	41.076	36.408	42.790	46.978
5	12.951	21.183	21.019	18.756	28.784	21.015	36.349	27.632	41.169	36.777	42.760	47.258
6	13.151	21.005	21.334	18.732	29.095	21.147	36.563	27.952	41.246	37.131	42.739	47.529
7	13.369	20.812	21.659	18.738	29.406	21.310	36.754	28.272	41.313	37.465	42.728	47.797
8	13.610	20.616	21.982	18.775	29.707	21.501	36.926	28.583	41.376	37.779	42.727	48.070
9	13.876	20.432	22.296	18.839	29.990	21.712	37.084	28.879	41.442	38.076	42.734	48.355
10	14.160	20.271	22.592	18.922	30.253	21.932	37.236	29.158	41.516	38.362	42.743	48.655
11	14.453	20.142	22.870	19.013	30.496	22.153	37.387	29.420	41.599	38.643	42.750	48.973
12	14.744	20.044	23.129	19.101	30.722	22.364	37.543	29.668	41.693	38.925	42.747	49.308
13	15.024	19.972	23.375	19.178	30.938	22.561	37.708	29.909	41.796	39.217	42.730	49.655
14	15.286	19.914	23.614	19.240	31.151	22.741	37.884	30.148	41.905	39.523	42.694	50.008
15	15.530	19.860	23.854	19.286	31.365	22.906	38.071	30.393	42.016	39.849	42.637	50.356
16	15.759	19.798	24.099	19.319	31.588	23.061	38.267	30.651	42.121	40.194	42.563	50.690
17	15.979	19.723	24.355	19.346	31.822	23.211	38.469	30.927	42.216	40.557	42.477	51.000
18	16.196	19.632	24.625	19.371	32.068	23.364	38.670	31.225	42.295	40.932	42.388	51.283
19	16.419	19.527	24.909	19.403	32.326	23.527	38.866	31.544	42.355 42.395	41.313 41.689	42.305	51.540
20	16.651	19.411	25.205	19.449	32.593	23.707	39.051	31.882	42.418	42.049	42.238	51.779
21	16.898	19.290	25.512	19.515	32.866	23.908	39.218	32.234	42.431	42.387	42.191	52.013
22	17.160	19.172	25.823	19.605	33.137	24.133	39.366	32.591	42.444	42.697	42.163	52.257
23	17.438	19.064	26.134	19.722	33.403	24.382	39.495	32.941	42.468	42.984	42.146	52.524
24	17.730	18.972	26.436	19.865	33.655	24.651	39.608	33.273	42.512	43.258	42.127	52.820
25	18.033	18.904	26.723	20.027	33.890	24.933	39.715	33.580	42.578	43.537	42.093	53.140
26	18.340	18.863	26.991	20.198	34.104	25.217	39.829	33.860	42.660	43.835	42.035	53.472
27	18.645	18.850	27.240	20.365	34.300	25.490	39.959	34.121	42.747	44.163	41.950	53.802
28	18.938	18.860	27.474	20.514	34.485	25.741	40.114	34.380	42.824	44.522	41.842	54.117
29	19.216	18.885	27.704	20.638	34.671	25.964	40.288	34.654	42.880	44.901	41.716	54.409
30	19.475	18.913			34.871	26.164	40.473	34.957	42.908	45.286	41.582	54.675
31	19.718	18.930			35.093	26.356	40.654	35.292	42.911	45.664	41.446	54.916
32	19.953	18.926			35.338	26.556			42.894	46.026		
	sec(δ)	tan(δ)	sec(δ)	tan(δ)	sec(δ)	tan(δ)	sec(δ)	tan(δ)	sec(δ)	tan(δ)	sec(δ)	tan(δ)
	10.51	10.46	10.51	10.46	10.51	10.46	10.51	10.47	10.52	10.47	10.52	10.48

Mean R.A. $15^h49^m33^s.396$ Double lower transit November 18 Mean Dec. $-84°32'33''.817$

APPARENT PLACES OF STARS, 2025

SOUTHERN CIRCUMPOLAR STARS AT UPPER TRANSIT AT GREENWICH

EQUINOX BASED RIGHT ASCENSION – WHOLE NUTATION

FK6 Star No. 1666 = Hipparcos Star No. 76996 = ρ Oct

Day	July		August		September		October		November		December	
	R.A.	Dec.	R.A.	Dec.	R.A.	Dec.	R.A.	Dec.	R.A.	Dec.	R.A.	Dec.
	15^h49^m	$-84°32'$	15^h49^m	$-84°33'$	15^h49^m	$-84°32'$	15^h49^m	$-84°32'$	15^h49^m	$-84°32'$	15^h49^m	$-84°32'$
1	41.446	54.916	36.978	00.790	30.967	61.872	25.537	57.976	22.464	49.961	23.468	40.781
2	41.316	55.136	36.814	00.870	30.793	61.842	25.381	57.815	22.394	49.648	23.580	40.438
3	41.194	55.342	36.658	00.959	30.610	61.822	25.213	57.645	22.339	49.308	23.722	40.102
4	41.083	55.541	36.506	01.061	30.414	61.807	25.035	57.455	22.310	48.950	23.889	39.792
5	40.983	55.740	36.352	01.178	30.201	61.786	24.854	57.235	22.313	48.588	24.068	39.517
6	40.892	55.947	36.191	01.309	29.973	61.748	24.681	56.982	22.343	48.242	24.244	39.277
7	40.806	56.166	36.015	01.448	29.735	61.682	24.529	56.703	22.392	47.925	24.404	39.061
8	40.720	56.400	35.821	01.587	29.499	61.583	24.405	56.410	22.443	47.639	24.543	38.852
9	40.628	56.651	35.608	01.715	29.275	61.452	24.309	56.122	22.483	47.379	24.663	38.637
10	40.524	56.915	35.381	01.821	29.072	61.299	24.234	55.852	22.506	47.130	24.771	38.404
11	40.403	57.186	35.147	01.896	28.894	61.140	24.166	55.609	22.510	46.879	24.876	38.151
12	40.262	57.453	34.916	01.940	28.738	60.989	24.095	55.391	22.499	46.613	24.985	37.877
13	40.102	57.706	34.698	01.956	28.595	60.858	24.009	55.187	22.482	46.327	25.106	37.587
14	39.928	57.935	34.500	01.956	28.454	60.749	23.906	54.987	22.464	46.019	25.242	37.289
15	39.750	58.135	34.324	01.955	28.304	60.659	23.786	54.777	22.455	45.690	25.396	36.990
16	39.576	58.306	34.164	01.965	28.138	60.579	23.653	54.550	22.459	45.344	25.568	36.698
17	39.415	58.455	34.012	01.995	27.953	60.498	23.515	54.299	22.479	44.990	25.756	36.419
18	39.274	58.593	33.858	02.047	27.751	60.402	23.379	54.023	22.519	44.634	25.955	36.158
19	39.153	58.735	33.691	02.116	27.536	60.285	23.252	53.724	22.577	44.284	26.161	35.919
20	39.045	58.892	33.504	02.193	27.315	60.141	23.139	53.407	22.651	43.946	26.369	35.702
21	38.941	59.073	33.295	02.263	27.098	59.970	23.044	53.081	22.735	43.625	26.573	35.506
22	38.829	59.277	33.067	02.316	26.889	59.774	22.966	52.752	22.826	43.323	26.768	35.323
23	38.699	59.498	32.827	02.343	26.694	59.562	22.906	52.428	22.918	43.041	26.953	35.148
24	38.545	59.722	32.582	02.341	26.516	59.339	22.860	52.114	23.005	42.773	27.126	34.972
25	38.366	59.935	32.340	02.312	26.355	59.114	22.823	51.816	23.084	42.516	27.291	34.786
26	38.169	60.128	32.107	02.260	26.207	58.894	22.790	51.534	23.153	42.262	27.454	34.585
27	37.960	60.292	31.887	02.193	26.071	58.684	22.756	51.268	23.212	42.002	27.621	34.365
28	37.748	60.428	31.681	02.118	25.941	58.488	22.717	51.012	23.266	41.728	27.803	34.128
29	37.539	60.540	31.490	02.043	25.812	58.306	22.668	50.763	23.320	41.434	28.007	33.882
30	37.340	60.632	31.310	01.974	25.679	58.137	22.608	50.511	23.384	41.117	28.238	33.640
31	37.153	60.713	31.137	01.916	25.537	57.976	22.538	50.246	23.468	40.781	28.494	33.416
32	36.978	60.790	30.967	01.872			22.464	49.961			28.767	33.223
	$\sec(\delta)$	$\tan(\delta)$	$\sec(\delta)$	$\tan(\delta)$	$\sec(\delta)$	$\tan(\delta)$	$\sec(\delta)$	$\tan(\delta)$	$\sec(\delta)$	$\tan(\delta)$	$\sec(\delta)$	$\tan(\delta)$
	10.53	10.48	10.53	10.48	10.53	10.48	10.53	10.48	10.52	10.47	10.52	10.47

Mean R.A. $15^h49^m33^s.396$ Double lower transit November 18 Mean Dec. $-84°32'33''.817$

REDUCTION TO HIPPARCOS AND FK6(LONG TERM) 2025
FOR MEAN EPOCH AND EQUINOX 2025.5

FK6-No.	HIP-No.	Hipparcos – FK6 (SI)				FK6: LTP – SI			
		$\Delta\alpha$	$\Delta\mu_\alpha$	$\Delta\delta$	$\Delta\mu_\delta$	$\Delta\alpha$	$\Delta\mu_\alpha$	$\Delta\delta$	$\Delta\mu_\delta$
10	1599	10.5	0.3	11.3	0.3	-16.4	-0.5	-10.7	-0.3
22	3419	-15.8	-0.5	6.2	0.2	2.0	0.0	-5.0	-0.1
44	5661	4.3	0.1	8.6	0.3	-6.2	-0.1	-13.1	-0.3
80	10642	43.0	1.3	-0.6	0.0	-3.9	-0.1	-0.5	0.1
110	14240	4.5	0.1	14.0	0.4	-20.7	-0.6	-17.7	-0.6
119	15510	-5.9	-0.2	0.2	0.0	20.6	0.4	4.1	0.0
120	15863	-1.5	0.0	-28.7	-0.8	-6.7	-0.2	8.3	0.2
153	19095	-5.6	-0.2	16.7	0.5	8.9	0.2	-0.1	-0.3
173	22361	27.9	0.8	20.1	0.6	-37.7	-0.8	-10.9	-0.3
201	25336	-13.1	-0.4	-37.8	-1.1	-10.2	-0.1	-5.5	0.1
203	25769	12.1	0.4	8.5	0.2	-5.6	-0.2	-2.6	-0.1
210	26311	2.6	0.1	15.6	0.4	0.3	0.0	-2.7	-0.1
223	27628	-1.8	-0.1	4.0	0.1	3.1	0.0	-12.1	-0.2
243	30324	11.0	0.3	-16.8	-0.5	-12.4	-0.2	5.3	0.1
245	30438	23.2	0.7	13.6	0.4	-2.1	-0.3	-4.4	-0.2
254	32246	-50.6	-1.5	-21.0	-0.6	-1.5	0.0	-4.1	-0.1
306	39429	-12.2	-0.4	13.0	0.4	9.2	0.3	-13.8	-0.4
323	42452	-6.8	-0.2	-32.1	-0.9	-1.8	-0.1	16.2	0.5
348	45238	79.9	2.3	27.4	0.8	-69.4	-1.6	-18.1	-0.7
354	46390	13.5	0.4	-15.8	-0.5	12.7	0.1	-10.3	-0.1
363	47594	4.7	0.1	21.7	0.6	11.1	0.2	-9.5	-0.3
422	54872	11.4	0.3	-17.6	-0.5	-21.8	-0.2	3.2	0.1
423	54879	31.0	0.9	-21.5	-0.6	-12.7	-0.2	0.4	0.1
426	55282	-16.5	-0.5	-8.6	-0.2	-5.5	0.0	3.2	0.1
455	59747	-8.8	-0.3	-11.7	-0.3	7.8	0.3	15.2	0.4
464	60823	-10.4	-0.3	-1.0	0.0	17.4	0.4	-5.3	-0.1
468	61084	-1.7	0.0	-11.2	-0.3	-10.3	-0.2	27.4	0.6
479	62131	-28.4	-0.8	3.2	0.1	12.4	0.4	14.6	0.4
509	67301	45.2	1.3	-44.6	-1.3	-17.5	-0.4	36.8	0.4
529	70069	13.3	0.4	-2.0	-0.1	-21.3	-0.7	3.6	0.1
560	74946	13.9	0.4	-6.5	-0.2	-41.4	-1.2	8.8	0.3
564	74785	28.1	0.8	-25.7	-0.7	10.8	0.1	5.2	0.1
588	77622	21.1	0.6	-5.2	-0.1	2.7	0.0	22.9	0.2
603	79593	-16.1	-0.5	30.6	0.9	-2.0	0.0	11.6	0.1
622	81377	22.5	0.7	-4.4	-0.1	-4.7	-0.1	-1.4	0.0
625	82273	-13.1	-0.4	24.5	0.7	4.8	0.2	-14.1	-0.4
635	83613	16.8	0.5	13.5	0.4	-12.8	-0.2	15.6	0.1
660	86670	1.0	0.0	-16.5	-0.5	-2.0	0.0	9.9	0.3
671	87585	25.7	0.7	14.3	0.4	-12.4	-0.3	-7.1	-0.2
691	90422	-0.3	0.0	9.1	0.3	-4.8	-0.1	-5.1	-0.4

Units: $0^{s}.0001$ for $\Delta\alpha$ — $0^{s}.0001/yr$ for $\Delta\mu_\alpha$

$0''.001$ for $\Delta\delta$ — $0''.001/yr$ for $\Delta\mu_\delta$

Reduction to HIPPARCOS and FK6 (Long-term prediction)

REDUCTION TO HIPPARCOS AND FK6(LONG TERM) 2025
FOR MEAN EPOCH AND EQUINOX 2025.5

FK6-No.	HIP-No.	Hipparcos − FK6 (SI)				FK6: LTP − SI			
		$\Delta\alpha$	$\Delta\mu_\alpha$	$\Delta\delta$	$\Delta\mu_\delta$	$\Delta\alpha$	$\Delta\mu_\alpha$	$\Delta\delta$	$\Delta\mu_\delta$
699	91262	-19.8	-0.6	27.4	0.8	-26.5	-0.2	-47.6	-0.4
700	90647	84.8	2.5	-26.6	-0.8	-38.4	-1.2	9.3	0.3
725	94834	-15.4	-0.4	-19.6	-0.6	-3.3	0.0	2.1	0.1
748	98495	2.3	0.1	9.2	0.3	3.4	-0.3	-12.4	-0.4
777	102098	-6.9	-0.2	14.9	0.4	1.1	0.0	-2.6	-0.1
855	112029	-2.9	-0.1	18.3	0.5	5.5	0.1	-7.7	-0.1
1039	6732	-11.2	-0.3	-6.5	-0.2	-4.5	0.0	-11.1	-0.1
1045	7513	-11.9	-0.4	14.6	0.4	-6.2	0.0	-7.8	-0.3
1116	19513	16.6	0.5	-28.3	-0.8	-4.7	-0.1	4.4	0.2
1166	29134	-1.9	-0.1	0.0	0.0	7.6	0.2	-1.6	0.0
1260	49339	4.4	0.1	54.0	1.6	4.2	0.0	-19.0	-0.5
1275	52098	7.7	0.2	-11.6	-0.3	-0.1	0.0	1.6	0.1
1307	57939	12.3	0.4	14.3	0.4	-21.0	-0.3	24.2	0.1
1357	67057	-6.7	-0.2	-7.9	-0.2	2.4	0.1	1.0	0.1
1396	73996	-4.2	-0.1	10.1	0.3	-5.9	-0.1	-27.8	-0.3
1456	84862	-4.1	-0.1	16.1	0.5	-2.9	0.0	-7.2	-0.2
1517	97290	0.0	0.0	-5.8	-0.2	-1.2	0.0	5.1	0.1
1533	101101	14.6	0.4	-15.0	-0.4	-5.7	-0.1	14.4	0.2
1546	102978	17.7	0.5	-43.9	-1.3	-4.3	-0.1	12.4	0.4
1577	108036	36.1	1.1	25.1	0.7	-3.8	-0.1	-14.3	-0.3
1644	72573	89.2	2.6	-7.5	-0.2	-48.7	-2.2	1.7	0.1
1662	30678	-134.8	-4.0	0.5	0.0	165.7	3.1	1.8	0.0
1666	76996	89.8	2.7	24.1	0.7	-152.7	-4.4	-16.0	-0.5

Units: $0\overset{s}{.}0001$ for $\Delta\alpha$ $\qquad$ $0\overset{s}{.}0001/yr$ for $\Delta\mu_\alpha$

$0\overset{\prime\prime}{.}001$ for $\Delta\delta$ $\qquad$ $0\overset{\prime\prime}{.}001/yr$ for $\Delta\mu_\delta$

Polaris (FK6 star No. 907) is not included in the list above because it is a double star. In the following table we give for Polaris the corrections from the apparent position based on the FK6 (see p. 26 and 27 of this publication) to the corresponding position based on the HIPPARCOS catalogue for the first day of each month. The corrections for intermediate days may be obtained by interpolation. The HIPPARCOS apparent place is obtained by adding the tabulated data to the FK6-position.

Reduction to HIPPARCOS for Polaris, 2025

Day	Month	Year	$\Delta\alpha[0\overset{s}{.}001]$	$\Delta\delta[0\overset{\prime\prime}{.}001]$
1	1	2025	736	85
1	2	2025	738	86
1	3	2025	736	87
1	4	2025	733	87
1	5	2025	731	87
1	6	2025	731	86
1	7	2025	731	86
1	8	2025	734	86
1	9	2025	740	84
1	10	2025	747	84
1	11	2025	754	83
1	12	2025	760	83
1	1	2026	764	83

TABLE UT – ST, 2025

SIDEREAL TIME AT 0^h U.T.

Date	Sidereal Time Apparent	Sidereal Time Mean	Equation of Equinox app–mean	Date	Sidereal Time Apparent	Sidereal Time Mean	Equation of Equinox app–mean
			(0.ˢ001)				(0.ˢ001)
	h m s	s			h m s	s	
Jan. 0	6 39 39.339	39.337	+ 2	Feb. 15	9 41 00.987	00.884	+ 102
1	6 43 35.905	35.893	+ 12	16	9 44 57.534	57.440	+ 95
2	6 47 32.468	32.448	+ 20	17	9 48 54.083	53.995	+ 88
3	6 51 29.028	29.003	+ 24	18	9 52 50.633	50.550	+ 82
4	6 55 25.584	25.559	+ 25	19	9 56 47.185	47.106	+ 79
5	6 59 22.137	22.114	+ 23	20	10 00 43.739	43.661	+ 78
6	7 03 18.688	18.669	+ 19	21	10 04 40.296	40.216	+ 80
7	7 07 15.240	15.225	+ 15	22	10 08 36.856	36.772	+ 84
8	7 11 11.794	11.780	+ 14	23	10 12 33.417	33.327	+ 90
9	7 15 08.352	08.336	+ 16	24	10 16 29.978	29.883	+ 96
10	7 19 04.914	04.891	+ 23	25	10 20 26.539	26.438	+ 101
11	7 23 01.478	01.446	+ 32	26	10 24 23.097	22.993	+ 103
12	7 26 58.045	58.002	+ 44	27	10 28 19.651	19.549	+ 102
13	7 30 54.612	54.557	+ 55	28	10 32 16.202	16.104	+ 98
14	7 34 51.177	51.112	+ 64	Mar. 1	10 36 12.749	12.659	+ 90
15	7 38 47.739	47.668	+ 71	2	10 40 09.296	09.215	+ 81
16	7 42 44.297	44.223	+ 74	3	10 44 05.844	05.770	+ 74
17	7 46 40.852	40.779	+ 73	4	10 48 02.395	02.326	+ 69
18	7 50 37.404	37.334	+ 70	5	10 51 58.950	58.881	+ 69
19	7 54 33.955	33.889	+ 66	6	10 55 55.509	55.436	+ 73
20	7 58 30.505	30.445	+ 61	7	10 59 52.070	51.992	+ 79
21	8 02 27.057	27.000	+ 57	8	11 03 48.632	48.547	+ 85
22	8 06 23.610	23.555	+ 55	9	11 07 45.193	45.102	+ 91
23	8 10 20.166	20.111	+ 55	10	11 11 41.752	41.658	+ 94
24	8 14 16.724	16.666	+ 58	11	11 15 38.307	38.213	+ 94
25	8 18 13.285	13.221	+ 63	12	11 19 34.859	34.768	+ 91
26	8 22 09.848	09.777	+ 71	13	11 23 31.408	31.324	+ 84
27	8 26 06.412	06.332	+ 80	14	11 27 27.955	27.879	+ 76
28	8 30 02.976	02.888	+ 88	15	11 31 24.502	24.435	+ 67
29	8 33 59.538	59.443	+ 95	16	11 35 21.048	20.990	+ 58
30	8 37 56.097	55.998	+ 99	17	11 39 17.596	17.545	+ 51
31	8 41 52.653	52.554	+ 99	18	11 43 14.146	14.101	+ 45
Feb. 1	8 45 49.204	49.109	+ 95	19	11 47 10.698	10.656	+ 42
2	8 49 45.754	45.664	+ 89	20	11 51 07.253	07.211	+ 42
3	8 53 42.303	42.220	+ 83	21	11 55 03.810	03.767	+ 44
4	8 57 38.854	38.775	+ 79	22	11 59 00.370	00.322	+ 47
5	9 01 35.408	35.331	+ 78	23	12 02 56.929	56.878	+ 52
6	9 05 31.967	31.886	+ 81	24	12 06 53.489	53.433	+ 56
7	9 09 28.528	28.441	+ 87	25	12 10 50.047	49.988	+ 59
8	9 13 25.092	24.997	+ 95	26	12 14 46.602	46.544	+ 58
9	9 17 21.656	21.552	+ 104	27	12 18 43.153	43.099	+ 55
10	9 21 18.218	18.107	+ 111	28	12 22 39.702	39.654	+ 47
11	9 25 14.778	14.663	+ 115	29	12 26 36.248	36.210	+ 38
12	9 29 11.335	11.218	+ 116	30	12 30 32.794	32.765	+ 29
13	9 33 07.888	07.773	+ 114	31	12 34 29.344	29.320	+ 23
14	9 37 04.438	04.329	+ 109	Apr. 1	12 38 25.898	25.876	+ 22
15	9 41 00.987	00.884	+ 102	2	12 42 22.456	22.431	+ 25

Table UT-ST – Sidereal Time at 0^h UT

TABLE UT – ST, 2025

SIDEREAL TIME AT 0^h U.T.

Date		Sidereal Time		Equation of Equinox		Date		Sidereal Time		Equation of Equinox	
		Apparent	Mean	app–mean				Apparent	Mean	app–mean	
				(0.ˢ001)						(0.ˢ001)	
		h m s	s					h m s	s		
Apr.	1	12 38 25.898	25.876	+	22	May	17	15 39 47.462	47.423	+	39
	2	12 42 22.456	22.431	+	25		18	15 43 44.024	43.978	+	46
	3	12 46 19.017	18.987	+	31		19	15 47 40.584	40.534	+	50
	4	12 50 15.580	15.542	+	38		20	15 51 37.141	37.089	+	52
	5	12 54 12.142	12.097	+	44		21	15 55 33.695	33.644	+	51
	6	12 58 08.701	08.653	+	49		22	15 59 30.246	30.200	+	47
	7	13 02 05.258	05.208	+	50		23	16 03 26.797	26.755	+	42
	8	13 06 01.811	01.763	+	48		24	16 07 23.347	23.310	+	37
	9	13 09 58.361	58.319	+	42		25	16 11 19.901	19.866	+	36
	10	13 13 54.909	54.874	+	35		26	16 15 16.460	16.421	+	39
	11	13 17 51.456	51.430	+	27		27	16 19 13.023	12.976	+	46
	12	13 21 48.003	47.985	+	19		28	16 23 09.590	09.532	+	58
	13	13 25 44.552	44.540	+	11		29	16 27 06.158	06.087	+	70
	14	13 29 41.102	41.096	+	6		30	16 31 02.724	02.643	+	82
	15	13 33 37.654	37.651	+	3		31	16 34 59.288	59.198	+	90
	16	13 37 34.210	34.206	+	3	June	1	16 38 55.848	55.753	+	95
	17	13 41 30.767	30.762	+	6		2	16 42 52.404	52.309	+	96
	18	13 45 27.327	27.317	+	10		3	16 46 48.958	48.864	+	94
	19	13 49 23.888	23.872	+	15		4	16 50 45.509	45.419	+	90
	20	13 53 20.448	20.428	+	20		5	16 54 42.060	41.975	+	86
	21	13 57 17.008	16.983	+	25		6	16 58 38.612	38.530	+	82
	22	14 01 13.565	13.539	+	26		7	17 02 35.165	35.086	+	80
	23	14 05 10.119	10.094	+	25		8	17 06 31.720	31.641	+	80
	24	14 09 06.670	06.649	+	21		9	17 10 28.278	28.196	+	82
	25	14 13 03.218	03.205	+	14		10	17 14 24.839	24.752	+	87
	26	14 16 59.766	59.760	+	6		11	17 18 21.402	21.307	+	95
	27	14 20 56.316	56.315	—	0		12	17 22 17.966	17.862	+	104
	28	14 24 52.869	52.871	−	1		13	17 26 14.531	14.418	+	113
	29	14 28 49.428	49.426	+	2		14	17 30 11.094	10.973	+	121
	30	14 32 45.991	45.982	+	9		15	17 34 07.656	07.528	+	128
May	1	14 36 42.556	42.537	+	19		16	17 38 04.215	04.084	+	131
	2	14 40 39.121	39.092	+	28		17	17 42 00.771	00.639	+	132
	3	14 44 35.684	35.648	+	36		18	17 45 57.324	57.195	+	129
	4	14 48 32.244	32.203	+	41		19	17 49 53.876	53.750	+	126
	5	14 52 28.800	28.758	+	42		20	17 53 50.427	50.305	+	122
	6	14 56 25.353	25.314	+	39		21	17 57 46.981	46.861	+	120
	7	15 00 21.904	21.869	+	35		22	18 01 43.538	43.416	+	122
	8	15 04 18.453	18.424	+	29		23	18 05 40.100	39.971	+	129
	9	15 08 15.002	14.980	+	23		24	18 09 36.666	36.527	+	139
	10	15 12 11.552	11.535	+	17		25	18 13 33.234	33.082	+	152
	11	15 16 08.104	08.091	+	14		26	18 17 29.802	29.638	+	165
	12	15 20 04.658	04.646	+	13		27	18 21 26.368	26.193	+	175
	13	15 24 01.215	01.201	+	14		28	18 25 22.930	22.748	+	182
	14	15 27 57.775	57.757	+	18		29	18 29 19.489	19.304	+	185
	15	15 31 54.336	54.312	+	24		30	18 33 16.043	15.859	+	184
	16	15 35 50.899	50.867	+	32	July	1	18 37 12.596	12.414	+	181
	17	15 39 47.462	47.423	+	39		2	18 41 09.147	08.970	+	177

TABLE UT – ST, 2025

SIDEREAL TIME AT 0^h U.T.

Date	Sidereal Time		Equation of Equinox
	Apparent	Mean	app–mean
	h m s	s	(0ˢ001)
July 1	18 37 12.596	12.414	+ 181
2	18 41 09.147	08.970	+ 177
3	18 45 05.699	05.525	+ 173
4	18 49 02.251	02.080	+ 171
5	18 52 58.806	58.636	+ 170
6	18 56 55.363	55.191	+ 172
7	19 00 51.923	51.747	+ 176
8	19 04 48.485	48.302	+ 183
9	19 08 45.049	44.857	+ 191
10	19 12 41.613	41.413	+ 200
11	19 16 38.177	37.968	+ 209
12	19 20 34.739	34.523	+ 216
13	19 24 31.298	31.079	+ 219
14	19 28 27.854	27.634	+ 220
15	19 32 24.407	24.190	+ 217
16	19 36 20.958	20.745	+ 213
17	19 40 17.508	17.300	+ 208
18	19 44 14.060	13.856	+ 205
19	19 48 10.616	10.411	+ 205
20	19 52 07.175	06.966	+ 209
21	19 56 03.738	03.522	+ 216
22	20 00 00.304	00.077	+ 227
23	20 03 56.871	56.632	+ 238
24	20 07 53.436	53.188	+ 248
25	20 11 49.998	49.743	+ 255
26	20 15 46.556	46.299	+ 258
27	20 19 43.111	42.854	+ 257
28	20 23 39.662	39.409	+ 253
29	20 27 36.212	35.965	+ 247
30	20 31 32.762	32.520	+ 242
31	20 35 29.312	29.075	+ 237
Aug. 1	20 39 25.865	25.631	+ 234
2	20 43 22.419	22.186	+ 233
3	20 47 18.976	18.742	+ 235
4	20 51 15.536	15.297	+ 239
5	20 55 12.097	11.852	+ 245
6	20 59 08.659	08.408	+ 252
7	21 03 05.222	04.963	+ 259
8	21 07 01.783	01.518	+ 264
9	21 10 58.341	58.074	+ 267
10	21 14 54.896	54.629	+ 267
11	21 18 51.447	51.184	+ 263
12	21 22 47.996	47.740	+ 257
13	21 26 44.545	44.295	+ 249
14	21 30 41.094	40.851	+ 243
15	21 34 37.646	37.406	+ 240
16	21 38 34.202	33.961	+ 241

Date	Sidereal Time		Equation of Equinox
	Apparent	Mean	app–mean
	h m s	s	(0ˢ001)
Aug. 16	21 38 34.202	33.961	+ 241
17	21 42 30.762	30.517	+ 245
18	21 46 27.325	27.072	+ 253
19	21 50 23.889	23.627	+ 261
20	21 54 20.452	20.183	+ 269
21	21 58 17.013	16.738	+ 274
22	22 02 13.570	13.294	+ 276
23	22 06 10.123	09.849	+ 274
24	22 10 06.673	06.404	+ 269
25	22 14 03.221	02.960	+ 261
26	22 17 59.768	59.515	+ 253
27	22 21 56.316	56.070	+ 246
28	22 25 52.865	52.626	+ 240
29	22 29 49.417	49.181	+ 236
30	22 33 45.971	45.736	+ 235
31	22 37 42.527	42.292	+ 236
Sept. 1	22 41 39.086	38.847	+ 239
2	22 45 35.646	35.403	+ 243
3	22 49 32.206	31.958	+ 248
4	22 53 28.765	28.513	+ 252
5	22 57 25.323	25.069	+ 254
6	23 01 21.877	21.624	+ 253
7	23 05 18.428	18.179	+ 249
8	23 09 14.976	14.735	+ 241
9	23 13 11.522	11.290	+ 232
10	23 17 08.069	07.846	+ 224
11	23 21 04.619	04.401	+ 218
12	23 25 01.172	00.956	+ 216
13	23 28 57.730	57.512	+ 218
14	23 32 54.291	54.067	+ 224
15	23 36 50.853	50.622	+ 231
16	23 40 47.415	47.178	+ 237
17	23 44 43.975	43.733	+ 241
18	23 48 40.531	40.289	+ 242
19	23 52 37.084	36.844	+ 240
20	23 56 33.633	33.399	+ 234
21	0 00 30.181	29.955	+ 226
22	0 04 26.727	26.510	+ 217
23	0 08 23.273	23.065	+ 208
24	0 12 19.821	19.621	+ 200
25	0 16 16.371	16.176	+ 195
26	0 20 12.923	12.731	+ 192
27	0 24 09.478	09.287	+ 191
28	0 28 06.035	05.842	+ 193
29	0 32 02.594	02.398	+ 196
30	0 35 59.153	58.953	+ 200
Oct. 1	0 39 55.712	55.508	+ 204

TABLE UT – ST, 2025

SIDEREAL TIME AT 0^h U.T.

Date	Sidereal Time		Equation of Equinox	Date	Sidereal Time		Equation of Equinox
	Apparent	Mean	app–mean		Apparent	Mean	app–mean
			$(0^s.001)$				$(0^s.001)$
	h m s	s			h m s	s	
Oct. 1	0 39 55.712	55.508	+ 204	Nov. 16	3 41 17.245	17.055	+ 190
2	0 43 52.270	52.064	+ 206	17	3 45 13.796	13.611	+ 185
3	0 47 48.825	48.619	+ 206	18	3 49 10.348	10.166	+ 182
4	0 51 45.377	45.174	+ 202	19	3 53 06.903	06.721	+ 182
5	0 55 41.926	41.730	+ 196	20	3 57 03.461	03.277	+ 184
6	0 59 38.472	38.285	+ 187	21	4 00 60.021	59.832	+ 189
7	1 03 35.019	34.841	+ 178	22	4 04 56.583	56.387	+ 195
8	1 07 31.567	31.396	+ 171	23	4 08 53.146	52.943	+ 203
9	1 11 28.120	27.951	+ 168	24	4 12 49.709	49.498	+ 211
10	1 15 24.677	24.507	+ 170	25	4 16 46.271	46.054	+ 218
11	1 19 21.238	21.062	+ 176	26	4 20 42.832	42.609	+ 223
12	1 23 17.801	17.617	+ 184	27	4 24 39.390	39.164	+ 225
13	1 27 14.364	14.173	+ 191	28	4 28 35.945	35.720	+ 225
14	1 31 10.925	10.728	+ 197	29	4 32 32.498	32.275	+ 222
15	1 35 07.483	07.283	+ 200	30	4 36 29.049	28.830	+ 218
16	1 39 04.038	03.839	+ 199	Dec. 1	4 40 25.600	25.386	+ 214
17	1 43 00.589	00.394	+ 194	2	4 44 22.153	21.941	+ 212
18	1 46 57.137	56.950	+ 188	3	4 48 18.711	18.497	+ 214
19	1 50 53.685	53.505	+ 180	4	4 52 15.273	15.052	+ 221
20	1 54 50.232	50.060	+ 172	5	4 56 11.840	11.607	+ 233
21	1 58 46.781	46.616	+ 165	6	5 00 08.410	08.163	+ 247
22	2 02 43.331	43.171	+ 160	7	5 04 04.978	04.718	+ 260
23	2 06 39.884	39.726	+ 158	8	5 08 01.545	01.273	+ 271
24	2 10 36.439	36.282	+ 158	9	5 11 58.107	57.829	+ 278
25	2 14 32.997	32.837	+ 160	10	5 15 54.665	54.384	+ 281
26	2 18 29.557	29.393	+ 164	11	5 19 51.219	50.939	+ 280
27	2 22 26.118	25.948	+ 170	12	5 23 47.772	47.495	+ 277
28	2 26 22.678	22.503	+ 175	13	5 27 44.324	44.050	+ 274
29	2 30 19.238	19.059	+ 179	14	5 31 40.876	40.606	+ 271
30	2 34 15.795	15.614	+ 181	15	5 35 37.430	37.161	+ 269
31	2 38 12.350	12.169	+ 181	16	5 39 33.986	33.716	+ 270
Nov. 1	2 42 08.902	08.725	+ 177	17	5 43 30.545	30.272	+ 273
2	2 46 05.451	05.280	+ 171	18	5 47 27.106	26.827	+ 279
3	2 50 02.000	01.835	+ 164	19	5 51 23.669	23.382	+ 286
4	2 53 58.549	58.391	+ 158	20	5 55 20.233	19.938	+ 295
5	2 57 55.102	54.946	+ 156	21	5 59 16.798	16.493	+ 305
6	3 01 51.659	51.502	+ 158	22	6 03 13.362	13.049	+ 313
7	3 05 48.222	48.057	+ 165	23	6 07 09.924	09.604	+ 320
8	3 09 44.787	44.612	+ 175	24	6 11 06.484	06.159	+ 324
9	3 13 41.354	41.168	+ 186	25	6 15 03.040	02.715	+ 326
10	3 17 37.919	37.723	+ 196	26	6 18 59.595	59.270	+ 325
11	3 21 34.481	34.278	+ 202	27	6 22 56.147	55.825	+ 322
12	3 25 31.039	30.834	+ 205	28	6 26 52.699	52.381	+ 318
13	3 29 27.593	27.389	+ 204	29	6 30 49.252	48.936	+ 316
14	3 33 24.145	23.945	+ 200	30	6 34 45.808	45.491	+ 317
15	3 37 20.695	20.500	+ 195	31	6 38 42.369	42.047	+ 322
16	3 41 17.245	17.055	+ 190	32	6 42 38.934	38.602	+ 332

Printing and Binding
Libri Plureos GmbH
Friedensallee 273, 22763 Hamburg, Germany